Making Human Geography

Also from Kevin R. Cox

Spaces of Globalization:
Reasserting the Power of the Local
Edited by Kevin R. Cox

MAKING HUMAN GEOGRAPHY

KEVIN R. COX

THE GUILFORD PRESS
New York   London

© 2014 The Guilford Press
A Division of Guilford Publications, Inc.
72 Spring Street, New York, NY 10012
www.guilford.com

Printed in the United States of America

This book is printed on acid-free paper.

Last digit is print number: 9   8   7   6   5   4   3   2   1

Library of Congress Cataloging-in-Publication Data is available
from the publisher.

ISBN 978-1-4625-1283-6 (paperback)
ISBN 978-1-4625-1289-8 (hardcover)

Preface

This is a book about the academic discipline of human geography. In particular, it looks at human geography from the standpoint of its theoretical development and its contributions to understanding concrete geographies. The title of the book conveys something of this, but it has, intentionally, multiple meanings. In the first place, there is the sense of human geographers constructing an interpretive framework—a framework for understanding the world. Human geography as an academic discipline has indeed been made by successive generations of people working as only academics do with quite abstract ideas: applying them in intellectual study and reworking that theory but always within a changing geohistorical context that has provided conditions for that (ongoing) work. It is, in short, a social construction. That is one sense in which the field is made. This is a crucial point of departure for my approach in this book.

Second, the title of the book is meant to express the history of that making. This is a more particularistic understanding of the making of human geography. As a university subject, geography in the anglophone countries dates only from the closing years of the 19th century. It then took at least another 50 years for human geography to make some sort of intellectual mark among the other human sciences. An awareness of theoretical and methodological debate came late to the field, and only after that could it take off as an intellectually serious field of inquiry.

Third, and finally, there is the making of human geographies themselves: of regions in their multiple relations to one another, of differentially scaled geographic spaces, of centers to peripheries, and of borders and boundaries. The important question here is the degree to which our understanding of that making has been enhanced by the changes in the field over the last 50 years. Most significantly, human geography discovered social theory, first in some of its more mainstream expressions, particularly from economics, and then more critical approaches, as in the interest in Marxism and subsequently in postmodernism, poststructuralism, and postcolonialism. Apart from postcolonialism, these were social theories that had always lacked a defensible spatial aspect, even while social life is always and necessarily spatial. One of human geography's self-imposed tasks since then has been to spatialize these theories: to demonstrate the difference that space makes once inserted as a necessary aspect of that theory. This has not only made a major contribution to a reworked social theory; once the spatial character of social relations is introduced, the effects are typically and utterly destabilizing. More significantly from the standpoint of human geography's more concrete goals, it has allowed a much more profound understanding of how actual geographies get reproduced and transformed.

This is a spatialization, moreover, which has been progressively applied to seemingly all facets of the social process: power, the division of labor, institutions, and the relation to nature and, latterly of course, to discourse. As this has occurred, I would argue, the interpretive abilities of human geographers have been radically transformed. New connections have been made between those different aspects, though always with the spatial in mind. One might even claim that, in its vanguard at least, human geography has achieved a holism that has eluded the other human sciences.[1] To understand a spatially differentiated world, this had to occur to a degree that it hasn't had to in the other human sciences.

The book grew out of the experience of teaching a graduate-level seminar in the history of geographic thought. This experience has now lasted for over 25 years and has given me the opportunity to reflect on the field and its development. As I already indicated, much has been accomplished, yet it seems to me that these accomplishments have occurred in spite of other more negative tendencies. There are two weaknesses I have identified in particular, and both suggest the importance of an enhanced awareness of how the field has developed.

The first is a certain sort of superficiality in the field. The slice of human geography's history of which most academic geographers are seemingly aware and which, more importantly, informs their approach and their practice, is remarkably thin.[2] One can be forgiven for knowing little about the discipline prior to the middle of the 20th century. But since then, the changes have been remarkable: one transforming wave after another, it might seem. What concerns me is a failure to absorb the lessons of the past and to assimilate the old to the new. There have been a series of "turns" in the field, starting with the spatial–quantitative work of the 1960s, working through behavioral and humanistic geography, Marxist geography, critical realism, the cultural turn, postmodernism, poststructuralism, and so on. But at each stage there has been an unnecessary rejection of too much. Ground has been vacated before it has been thoroughly turned over and cultivated. Each "turn" has reacted to what has gone immediately before it and, possibly to justify itself, has been too sweeping in its rejection. This is not the continuity amidst revolution that Peter Taylor (1976) claimed in his quantification debate article as the dominant and necessary theme in the history of scientific fields. There *are* continuities. Much that we take to be new made an appearance earlier. Hägerstrand and Curry were talking about chance juxtapositions and their roles in the transformation of particular human geographies long before the current interest. But what we remember from the past often tends to be garbled and to reflect the self-justifying claims of the different "turns" as they have tried to define themselves against their predecessors.[3]

The second conclusion is that of a highly fragmented field. I am not referring here to the fragmentation along systematic lines that has been bemoaned by many. On the contrary, and as I remarked previously, an important current in human geography has brought the cultural, the social, the economic, the ecological, and the political increasingly together. Rather, it is division by theory and method that is my concern. The general disdain for the legacy of the spatial–quantitative revolution is particularly bothersome. This is because human geographers could benefit immensely not just from a knowledge of the extraordinary intellectual legacy of the spatial–quantitative revolution, in particular the way it set the scene for the development of the field in the remainder of the century, but also from some elementary awareness of how measurement can complement their own approaches. Human geographers talk a great deal about the geography of uneven development but have only the crudest grasp of how it is configured and how it is changing. There are also the critical human geographers who only loosely and uncomfortably overlap with the Marxist geographers, and there are those most strongly influenced by the "posts" who laugh the Marxist geographers out of court because of their "essentialism," "reductionism," "materialism," "economism," or some other knee-jerk reaction, a conclusion that is arrived at with little or no reading of the foundational—and the word is deliberately chosen—texts.

My own position is that of someone who has experienced during his academic career all the different turns that human geography has made over the last 50 years. After an undergraduate education that still bore the traces of the people–nature and regional work of the 1950s, I was initially a spatial–quantitative geographer, then a behavioral geographer, which was followed by a less structured interest in the social relevance movement. This was succeeded by an interest in historical materialism and critical realism and then more exclusively the former. I have experienced many of the debates firsthand and often with mixed feelings. It has also led me to my current views as to where we should be going, and these will become evident as the reader progresses through the book. I would say, right at the start, though, that I remain convinced that there is a shared history, a shared body of concepts, and a set of debates of interest and importance to all human geographers. I want to convey that shared knowledge.

Finally, and before turning to the body of the text, there are some aspects of my approach of which the reader should be aware. The first point is that this book is almost exclusively about anglophone geography, with some reference to those outside this sphere, like Hägerstrand, who participated in it. To have included French and German geography would have made for a very different book, even if I had had the knowledge to do it. In any case, my impression is that anglophone geography has been much more dynamic in its development than either of its continental cousins; perhaps there is both more to say and more to celebrate in a book that focuses exclusively on the first.

A second point concerns emphasis. Very crudely, books are either expository, as in "how to do it," or interpretive: a critical scrutiny and reworking of what people have written. This book is unusual in that it is both. With respect to the expository, I have felt it important to provide the reader with exemplars through which particular points in the argument can be more thoroughly grasped. It is not enough to say that quantitative geography has moved away in the last 20 years or so from its earlier generalizing mode in the direction of incorporating contextual considerations. Rather, it is important to know exactly what that means in terms of procedures. Likewise, one can argue that the resolution of the structure–agency debate at the beginning of the 1980s had little effect on research practice. But, whether or not that formal resolution was responsible, research in the rest of the decade quite clearly reflected an understanding of what was at stake. Just how that was expressed again requires reference to particular instances at some length. It is possible that at times some will find this unnecessary. But if they take to heart a concern at the core of this book that we have rejected too much in the past, then it will only be "at times," and they can skip the material that they feel is already familiar.

The fact that the book's approach is emphatically interpretive should also come with a caveat. I want to convey an understanding of why things happened the way they did; the implications of particular developments for the field as a whole; the pros and cons of the various debates; and the position of human geography with respect to social theory as a whole and how it has clarified real geographies both on the ground and in the

mind. This is a personal understanding: It is rooted in my own intellectual history, though others have contributed significantly to that history
and, I am confident, would support many of the points I make. It means,
however, that the work of some is more copiously referred to than that of
others. In part, but only in part, this is because in my view they have had
more effect on the discipline's trajectory. Other citations are there more
by chance: I happened to have read them at some time in the past and
then noted their particular power in demonstrating a claim that needed
to be made at a particular point in my argument. One result of this is
that the book's bibliographical references are by no means confined to
the people who typically get the most citations or are the most visible on
other counts. In any case, whether they are cited or not, it is the human
geographers of the last century, and particularly of the last half-century,
who have been the inspiration for this book.

NOTES

1. The collection *Geographies of Economies*, edited by Roger Lee and Jane Wills
   (1997), was a landmark in this regard.
2. An example from my seminar: In 2008 on the first day of the seminar, I sam-
   pled awareness of major 20th-century geographers. Of 12 PhD students taking the seminar, seven had heard of Doreen Massey, five of Michael Watts,
   but only three of Hägerstrand, two of Brian Berry, one each of Bill Bunge
   and Peter Gould, and, amazingly, none had ever come across Bill Garrison. It
   may be argued that graduate students are less likely than faculty to be aware.
   On the other hand, and if this is the case, why weren't faculty communicating
   an awareness of these names, which are surely part of human geography's
   pantheon?
3. I have long struggled to convey to students that the "spatial" part of the
   spatial–quantitative revolution had important antecedents. But, of course,
   one of the ways in which the spatial–quantitative geographers sought to pur-
   sue their agenda was through emphasizing the utter novelty of their project.

Contents

CHAPTER 1

Human Geography: The First Half Century

This is a book about human geography as an academic field: as a discipline taught in schools and represented in the universities as an accepted area of study. As such, it dates from approximately the turn of the last century. The Schools of Geography at Oxford and Cambridge were founded in 1899 and 1889,[1] respectively. In the United States, Departments of Geography were established at the University of Chicago in 1903, Harvard in 1885, and at the University of California at Berkeley in 1898.[2]

What I try to do in this chapter is to provide a survey of what human geography was like over the first approximately 50 years of its existence in the universities of the anglophone world. From the mid-1950s on there is clear evidence of accelerating change, and I argue that the second half of the 20th century was quite different from the first. For the first time, theory entered the geographer's vocabulary, and methods became something to be put in question and to be improved on. Many of the relationships that had been the focus of the human geographer's concern remained the same but were now looked at in a very different light. The same applies to the concepts through which those relations were grasped: change amidst continuity.

This is not to claim that the first half century can be described in a monotone. There is a history. The particular relations that human geographers placed at the center of their concerns and how they viewed those relations tended to shift as the century progressed. In his 1973 Presidential Address to the Association of American Geographers (AAG), Edward

1

Taaffe tried to capture those shifting foci in terms of what he called three traditions: area studies, man–land, and the spatial. I draw on this characterization here, demonstrating that there was indeed a clear history. I also show that there was some conceptual development even while it was not argued out in terms any better than an appeal to the facts of the case. The human geography of the time was nothing if not positivistic, and this was to last until the beginning of the 1970s. But neither was it static.

PEOPLE AND NATURE DOMINANT

Any statement is of geographical quality if it contains a reasonable relation between some inorganic element of the earth on which we live, acting as a control, and some element of the existence, or growth, or behavior, or distribution of the earth's organic inhabitants, serving as a response.

—DAVIS (1906, p. 71)

The first half of this century has seen the emergence of the modern study of geography as an academic discipline fit to take its place among the older disciplines of science, the social sciences, and the liberal arts in every university of Britain. This remarkable and rapid growth, paralleled by that of the study of geography at all levels in schools, is undoubtedly due to the realization that there is an intimate relationship between man and his environment and that no other subject seeks to understand or interpret this relationship in its entirety both in space and time.

—STAMP (1960, p. 9)

In human geography in the first part of the last century, the man–land tradition as Taaffe referred to it was utterly dominant. This was the human geography that was still being taught in British high schools and universities in the 1950s, as per the second quote just presented from one of the major figures in British geography at that time, Dudley Stamp. There were, nevertheless, important shifts in emphasis. Initially it was the role of the physical environment in structuring human activities that was stressed. Later there would be reactions to this. One emphasized the role of culture in changing the physical environment. A second saw the relation to the natural environment as one that was mediated by culture and technology: environment as a technical and cultural appraisal, therefore. Under all these headings, there are literatures of varying degrees of methodological sophistication and theoretical penetration. All of them, though, relied on a rather stark, and in some ways indefensible, separation of the natural from the human. This had important implications for theory in human geography, even while "theory" in an explicit, acknowledged sense was still *terra incognita*. In this section I start with a brief history of work in this vein during the first half of the last century and then discuss some of its implications.

A Mini-History

Environmental Determinism

From the late 19th century to the 1930s and with lingering influences well beyond, human geography was dominated by what was known as environmentalism. The main concern was documenting how the natural environment influenced, determined, controlled, or conditioned human geographies, including esoteric geographies such as "civilization." The names of people like Ellsworth Huntington, Ellen Churchill Semple, and James Fairgrieve are notable here, among others. A first example comes from British geographer Halford Mackinder. In *Britain and the British Seas*, published in 1907, Mackinder made the following claims about what he called "the essential qualities of the British environment":

(1) Insularity, which has tended to preserve the continuity of social organization;

(2) Accessibility, which has admitted stimulus from without, and prevented stagnation;

(3) Division into a more accessible east and a less accessible west, which has made for variety of initiative, and resulting interaction;

(4) Productivity of soil and climate, the necessary basis of a virile native growth;

(5) Possession of a vast potential energy stored in deposits of coal, the mainspring of modern industrial life; and

(6) Interpenetration by arms of tidal sea, giving access to the universal ocean-road of modern commerce. (pp. 178–179)

This is typical of this genre of writing: the emphasis on the effects of the natural environment and, not least, the failure to provide any serious corroboration for the claims. More interesting in this regard was the work of another of the determinists: American geographer Ellsworth Huntington.

Huntington's obsession was "the geography of civilization" and the role of climate in producing it.[3] His argument was that climate influenced health and energy, and this in turn influenced the development of what he called "civilization." The latter was something to be measured. On a world scale, it was a matter of asking a group of "experts" to rank countries in terms of their respective levels of civilization. In examining his thesis within the United States, he drew on a number of different measures, including white homicide rates as an (inverse) measure of "self-control" and various social and economic variables as indicators of "the goodness of life." These measures were correlated using maps with measures of what he called "climatic efficiency," drawing on levels of productivity of (1) factory workers and students around the world (Huntington, *Civilization and Climate*, 1915, pp. 228 and 234) and (2) piece workers in the United States (Huntington, *The Mainsprings of Civilization*, 1945, Chap. 12). The different measures were mapped, and "tests" of his hypotheses were then in the form of copious map comparisons.

In mitigation, Huntington clearly did not regard climate as the only significant variable; rather, its importance was, at least in his intentions if not execution, to be evaluated relative to that of other variables, notably race. Accordingly, he believed that "low races" could become more civilized if transplanted out of their "original, unstimulating environments" (1915, p. 59).[4] There is also occasional reference to social conditions: the role of public opinion in encouraging self-discipline in the face of the challenge of the various pathologies of alcoholism, laziness, and immorality, which he believed were brought on by tropical climatic conditions. Likewise and in anticipation of a direction that human geography would later take in response to the excesses of environmental determinism, there was the occasional nod in the direction of the difference that technology might make to the climate–"civilization" relationship.[5] In all these regards, Huntington was a clear advance on Mackinder. What they shared, though, was a distinctive intellectual environment.

Livingstone has argued that studies of climate and its relation to people in the late 19th and early 20th centuries of the sort exemplified by Huntington's writings were permeated by political and moral evaluation: Peoples were evaluated, found wanting, lacking civilization, and so on, and their moral defects explained in terms of climate. This was often with the political subtext of why such peoples not only were colonized peoples but why they *should* be. It also reflected an interest in the possibilities of white settlement in the tropics and the problem of "acclimatization."

This may be. It does not help shed light on the quotes from Mackinder, though, nor the writings of Huntington. Rather, I think there was a genuine attempt to try to understand the particularities of development in the world or what was known at that time as "civilization" through appeal to geography and so to justify its intellectual importance if only in the eyes of its practitioners. One is struck here by Huntington's appeal to a diversity of geographical determinants, including race, even while he tended to put almost all his emphasis on climatic variation. In this regard, Livingstone's (1992) argument that the environmental focus was part of an effort to reconstitute and professionalize geography in the face of the academic atomization that was proceeding apace at the turn of the century seems thoroughly apropos.

Possibilism

In its original, more primitive form, the man–land approach had emphasized the effects of the natural environment on geographic differences, particularly of development and "civilization." An alternative approach reversed the causal arrow between "man" and the "land." Instead of natural limits, controls, or conditions, one now talked about the role of culture in changing those limits or in making choices among a diverse set of natural conditions. The emphasis shifted to human action as causative and on choice as conditioning human action, although in some cases the effects of those choices might be thoroughly unintended.

An early expression of this is the work of French geographer Vidal de la Blache, who argued for a "scientific study of places" in which what he called "genres de vie," or modes of living, would be placed at the center of the picture. These were place- or region-specific ways of life, and he sought to explain them in terms of particular articulations of the natural milieu and people's interpretation and mobilization of the possibilities latent within that milieu. This early reversal of the causal arrow was known as "possibilism" to emphasize its rejection of the physically determinist nature of environmentalism.

A more developed form of this argument took shape in the thinking of American geographer Carl Sauer, who was particularly interested in the human impact, intended and unintended, on the natural environment. Sauer figured prominently in W. L. Thomas's critical collection of essays, *Man's Role in Changing the Face of the Earth*. Specific research foci for Sauer (1967) included the role of fire in replacing woodland with grassland. Burning was used to increase the yield of desired animals and plants. This also changed the nature of the trees that survived: those that were more fire resistant, those that germinated and grew quickly after a fire and those that could tolerate full exposure to sun. According to Sauer, "The climatic origin of grasslands rests on a poorly founded hypothesis." The same conclusion applied to some deserts. As far as the deserts of the southwestern United States were concerned, there was historical evidence of the land bearing large numbers of cattle in the 18th century. Sauer argued that under those conditions each successive drought left the range depleted, its carrying capacity reduced, and the recovery of the range less likely.

In the United States Sauer's influence was substantial. In Britain it was less evident, and most of the topics Sauer was interested in fell to the systematic field of historical geography. Excellent examples include H. C. Darby's monograph on the draining of the Fenland (1956)[6] and Clifford Smith's work on the origin of the Norfolk Broads, a series of small, shallow lakes that were eventually traced to medieval peat diggings (Lambert, Jennings, Smith, and Godwin, 1960). Sauer's interest in premodern landscapes is confirmed here.[7]

The Natural Environment as a Technical and Cultural Appraisal

A less explored theme in the history of human geography's early love affair with the relations between people and "their" natural environment was how that relation was mediated by technology and social values. There is certainly an overlap here with Carl Sauer, but it not only predates him; it was also less anxious about demonstrating how human activity modified the environment. One could certainly acknowledge that geology, climate, and so on might play some role in understanding a geographic distribution. It was, however, only in terms of the technologies, the understandings of nature, and people's values that this could make sense.

This line of reasoning was evident as early as the mid-1920s in the work of Daryll Forde (1925, 1934), a British geographer who, appropriately enough given the nature of his arguments, had strong ethnological interests and would later migrate to anthropology. In his 1925 paper with the significant title "Values in Human Geography," Forde was at pains to point out the way in which a variety of practices ranging from clothing habits to agricultural concentrations could not be reduced to climatic variation. He was also keenly alert to the implications of technological development, remarking on among other things, the significance of the development of rapid oceanic navigation and refrigeration for the (then recent) agricultural development of countries like Argentina and New Zealand. On the other hand, and certainly a reflection of the intellectual environment of the time, he saw technology as an expression of a particular stage of development or civilization.

The notion of civilization as an explanatory condition would continue into the postwar period. A book that had a major impact on British geography at least during the 1950s was *The Tropical World*, by French geographer Pierre Gourou (1953). Much of its appeal stemmed from Gourou's careful examination of the nature of tropical soils and what they implied for human adaptation and population densities. The rapid exhaustion of the soil subsequent to the clearing of the forest resulted in shifting forms of cultivation over most of the equatorial zone. Yet there were clearly exceptions where, despite these challenges, relatively high rural population densities were achieved. The answer, he claimed, was differences in civilization. Monsoon Asia was a case in point, since it "taught us that low population densities are not a necessary result of tropical conditions. The tropical environment certainly provides many obstacles, but these can be overcome; the vast areas of dense population in tropical Asia contain peoples with a well-developed civilization, whilst the sparsely populated areas of the tropical world are occupied by civilizations whose techniques of production and political organization are rudimentary" (p. 140).

Where the focus was difference within the developed world, then technological change would suffice as the explanatory fulcrum. By the 1950s, this had become a common theme in human geography research, particularly among historical geographers. The historical chapters of Wilfred Smith's (1949) *An Economic Geography of Great Britain* provide quite startling exemplars of this as the author traces out the implications of new forms of technology in the production and distribution of energy for the country's changing population geography: from textile mills on running water, to heavy industry on the country's coalfields with the invention of the steam engine, to the later dispersion of light industry courtesy of electricity. Another instance among many is Howard G. Roepke's 1956 study of the changing geography of the British iron and steel industry, in which, for example, the changeover from charcoal smelting to coke smelting led to a shift from the woodlands to the coalfields and the invention of the Gilchrist–Thomas basic hearth steel process in the late 19th century opened up to exploitation the iron ores around Middlesborough.

Intriguingly, a more general and programmatic statement of the people–environment relation along these particular lines would only come very later on—in fact, after the onset of the spatial–quantitative revolution. This was William Kirk's (1963) "Problems in Geography." Although Kirk was primarily concerned with identifying a rationale for the field of geography as a whole, his development of the idea that it should be "the environment as a field of action" had, in his hands, important implications. As such, it was to be understood in terms of human consciousness: "a psycho-physical field in which phenomenal facts are arranged into patterns or structures and acquire values in cultural contexts" (p. 366). This was the behavioral as opposed to phenomenal environment, and he clearly saw the former as incorporating the technological.[8] In this way, he hoped to short-circuit the idea of the environment as a thing apart, though without moving beyond the sort of emphasis on communities as agents that Sauer had espoused—a crucial weakness, as would become clearer when human geography later embraced social theory in a self-conscious way.

Some Summary Comments

What needs to be emphasized above all is the utter pervasiveness of the man–land tradition in Anglo-American geography during the first half of the 20th century and extending even into the 1950s. The vapidities of environmental determinism were by then long gone, but the idea of an orderly relationship between people and their environment as the dominant one to be considered in evaluating human geographies persisted.

The reasons for this are buried deep in academic geography's beginnings, in part the way it was often first nourished within geological departments but, and perhaps more important, as a result of the massive influence that Darwinian thinking exercised over thought, including social thought in the second half of the 19th century. Darwin's emphasis was clearly biological: Traits were selected in as a result of the way they facilitated survival in a particular environment. Human beings, on the other hand, could, by virtue of their cultural capacities, adapt in ways that were technical and cultural. The important point is that Darwin lived on in human geography through a concern for the adaptation of organisms to their natural or physical environments. The initial move in this direction was from the environment to the organism: Nature imposed a particular way of life on people by virtue of the incentives or disincentives that it provided. Later attention moved to the distinctive characteristics of people as biological organisms.

Accordingly, the people–environment, or "man–land" relationship as it was called, would continue to be *the* criterion of significance in geographical description and understanding. Geography was the natural environment, as reflected in books with titles like *The Geography Behind History*. When carried to its logical conclusion, the results could be quite bizarre, though at the time they seemed perfectly reasonable. Kenneth Sealy's 1957 *The Geography of Air Transport* provides a case in point. One rapidly gets the idea. After an Introduction, his Chapter 2 has the ominous title "The Physical Geography of Aviation." We learn, among other things, that "high mountains are hazards" (p. 34) while "forested zones, especially those within the tropics or the northern taiga present little real obstruction. Forced descent may be a hazardous business, but the presence of lakes and rivers mitigates these perils" (p. 35).[9,10]

The emphasis on the relation between people and nature meant a highly eviscerated sense of the social, if indeed it existed at all. One might recognize that the natural environment was a technological and cultural appraisal without it resulting in any reflection on the social conditions within which particular technologies or cultural values might develop. People, in other words, were assumed to be people: not much more than biological organisms with certain needs for food and water. So it is not surprising that one of the subthemes in human geography both before and after World War II was the relation between population and resources. This was of interest to Sauer himself:

> The steeply increasing production of late years is due only in part to better recovery, more efficient use of energy, and substitution of abundant for scarce materials. Mainly we have been learning how to deplete more rapidly the resources known to be accessible to us. Must we not admit that very much of what we call production is extraction?

> Even the so-called "renewable resources" are not being renewed. Despite better utilization and substitution, timber growth is falling farther behind use and loss . . . Much of the world is in a state of wood famine, without known means of remedy or substitution. (1967, pp. 24–25)

This was also something picked up on in the political geography of the time. In *The New Europe*, Fitzgerald (1945) talked about the need for a redistribution of population between the countries of the world.[11] Italy's major problem was the land hunger of its peasantry (p. 202). On the other hand, in light of heightened imperial tensions, too few people could be a problem. For Fitzgerald, therefore, one of France's major problems was that it was "deficient in population." There is a similar emphasis in a later compilation entitled *The Changing World: Studies in Political Geography*, edited by W. Gordon East and A. E. Moodie (1956).

There were exceptions to this disinterest in the social. The British geographer Mackinder and the American Bowman, who were writing prior to 1930, were unusually sensitive to issues of class relations, as we see later in a discussion of their work in Chapter 9. Both also had a heightened sense of human agency. They were well aware of the role that governments played in creating human geographies, but this was never something that, like Sauer's culture, was a sort of reified force over which people had no control. In fact, both of them wrote because they believed that they could make a difference. Significantly, both were men of action themselves as well as people with a public agenda.

Mackinder also had on occasion a quite developed sense of the significance of social relations in the abstract. It wasn't something that simply informed his policy agenda. In this regard, he is in vivid contrast to the dominant man–land view of the time: a view in which social relations, as we have seen, had little place. In the early part of *Democratic Ideals and Reality* (1919) there are some striking statements along these lines:

> The modern reality of human control over nature, apart from which democratic ideals would be futile, is not wholly due to the advance of scientific knowledge and invention. The greater control which man now wields is conditional, and not absolute like the control of nature over man by famine and pestilence. Human riches and comparative security are based today on the division and co-ordination of labor, and on the constant repair of the complicated plant which has replaced the simply tools of primitive society. In other words, the output of modern wealth is conditional on the maintenance of our social organization and capital. (1919, pp. 10–11)

And:

> For every advance in the application of science there has been a corresponding change in social organization. It was by no mere coincidence that Adam

Smith was discussing the division of labor when James Watt was inventing the steam engine. Nor, in our own time, is it by blind coincidence that beside the invention of the internal combustion engine—the key to the motor car, submarine and aeroplane—must be placed an unparalleled extension of the credit system. (1919, p. 12)

These are quite remarkable statements. The relation to nature is affirmed, but now it is socially mediated. Likewise, and as per the notion that nature is a technological appraisal, we are put on clear notice that this needs to be understood in the context of social relations, in this particular instance the division of labor and the credit system. It would be a long time before human geographers thought along those lines again.[12] In the meantime, the recipes of the man–land tradition persisted. Among other things, they found application in the study of particular places or regions. It is to that topic that we now turn.

The Area Studies Tradition

The idea of difference between places, ordered or otherwise, has always been central to the practice of geography as an intellectual pursuit. One of the earliest of those commonly recognized as geographers, Strabo, the ancient Greek scholar, divided the world into three zones: the torrid, the temperate, and the frigid. Generations of British schoolchildren were inducted into a division of the country into upland and lowland Britain, typically separated from one another by a line drawn from the Exe to the Tees or, alternatively, from the Bristol Channel to the Wash. This parallels more lay understandings of geography as in the common attribution of regional labels: the South, the Midwest, and so on.

There has been considerable variation in the way this interest has been expressed. On top of that, the tradition has fluctuated in its degree of centrality to the academic geography enterprise. Early in the 20th century, the idea of the region did indeed command attention. Typically seen as qualitatively distinct and singular, it came to displace environmentalism as the core of American geography. The conditions for this were at least threefold. One was simple disillusion with the pseudoscience of environmentalism represented by its more extravagant claims and the seeming elusiveness of coming to conclusions about, for example, influences, determinants, and controls (i.e., just *how* controlling/influencing?). A second was the search for an object that human geography could call its own, in the manner of the other sciences, while the third were the well-worn tracks of German and French geographers, including Vidal de la Blache. All these influences came together in the work of the great American geographer and founder of the so-called Berkeley School, Carl Sauer. Sauer's magisterial statement is "The Morphology of Landscape" (1925).

In that lengthy article, he expresses disillusion with environmentalism; embraces the idea that every science has to have a phenomenon that it can call its own; and then, under the clear influence of German and French geographers from the first two decades of the century, identifies what that object should be: the cultural landscape or culture area.

In all these attempts, the human–environment distinction was retained as a key one, but whereas some, like Hettner, emphasized "the physical basis," others, including Schlüter, stressed the transformative role of human agency. It was the latter to which, notably, Sauer was attracted. The manner in which he talked about it, though, underlines his adherence to a nature–culture dichotomy: "The cultural landscape is fashioned out of a natural landscape by a culture group. Culture is the agent, the natural area is the medium, the cultural landscape the result" (1925, p. 25.)

Several features of Sauer's concept of the cultural landscape are notable. The first was the focus on material change in the landscape and the material expressions of human culture: "The cultural landscape is the geographic area in the final meaning. Its forms are all the works of man that characterize the landscape. Under this definition we are not concerned in geography with the energy, customs or beliefs of man but with man's record upon the landscape" (1925, p. 25). The second is the focus on human transformation of so-called natural landscapes, so preserving the natural–cultural or nature–human dichotomy. According to Sauer, "The cultural landscape is fashioned out of a natural landscape by a culture group. Culture is the agent, the natural area is the medium, the cultural landscape the result" (1925, p. 25) Finally, cultural landscapes were organic wholes. They consisted of elements—farmhouses, property lines, particular land use complexes, field boundaries, patterns of land use— that were necessarily related with respect both to each other and to the natural conditions (e.g., the underlying geology and building materials). Sauer quotes approvingly the statement that one has not fully understood the nature of an area until one "has learned to see it as an organic unit, to comprehend land and life in terms of each other," which, of course, resonated closely with Vidal de la Blache's view.

This was a very particular view of the region and one worth dwelling on at some length. A clue is provided by Tony Wrigley (1965) in his critique of the Vidalian method, and it applies equally to Sauer. Wrigley pointed out that Vidal de la Blache's method was fine for traditional societies but broke down when one tried to apply it to societies that fulfilled their material needs not necessarily locally but through quite elongated links of an exchange nature. The importation of building materials like brick could result in the displacement of the locally available but more expensive limestone or sandstone for house building. But neither Vidal de la Blache nor Sauer were interested in contemporary, urban, societies and perhaps this is the reason.

Their view of the region was significantly antimodernist. It led to a focus on what would later be called "formal regions" or regions of homogeneity[13] and for which, in their cases, relations with other regions, perhaps through some geographic division of labor, were immaterial to their character. Areas characterized by some self-sufficiency as well as some unity of cultural landscape and form—in other words the sorts of regions that, as Wrigley implied, were rapidly disappearing—were of particular interest.[14] These were by no means idiosyncratic views. Other luminaries of the regional geography of the period, like Herbert Fleure and E. Estyn Evans, had similar biases.[15] This approach might work when applied to Ireland's Celtic Fringe, the Welsh Uplands, or more isolated parts of France like the Basses Alpes, but areas like that were becoming fewer and farther between. In the United States this was particularly the case. Applying the method to the rich corn lands of East Central Illinois just did not work. Perhaps significantly the only regional study that Sauer carried out in the United States was of the Ozarks.

It might seem, therefore, that the way in which regional geography was practiced was not at all in tune with the times. This is both right and wrong. It is right in the sense that people's lives were increasingly commodified and urbanized. The sort of self-sufficient peasantry selling only its surplus had long disappeared in England and the United States and was fast disappearing in the rest of Western Europe. Even if people lived in the countryside, they might well work and shop in a neighboring town, and farmers would certainly market their grain and livestock there. Towns and cities were increasingly the points from which social life was organized. In these regards, the sort of region favored by the likes of Sauer, Fleure, and Evans had rapidly diminishing significance. But there were countervailing forces. In the 1930s in particular, the countryside and the land were valorized, and this intersected with a longer standing antiurbanism, part of which is evident in Mackinder's anxieties about centralization in London and the importance of what he called "balanced communities" (Mackinder, 1919, Chap. 7). This valorization reached its climax in Nazi Germany. It was, however, much more widespread than that.[16] It also intersected with demographic anxieties of urban degeneration and the moral virtues of working the land, which had a longer history[17] and with which the emphasis on the relation to nature of contemporary geographers found a nice convergence.

I should note two other factors regarding the regional geography of the time. The first is how few exemplars there are compared with, for example, French geography and its tradition of regional monographs: the work of not just Vidal de la Blache, therefore, but, and among others, Albert Demangeon, Max Sorre, and Roger Dion. Rather, the region seems to have been a descriptive tool for engaging with the geography of a larger area, typically a country. A geography of the United States would include at some point a division into regions, typically along the formal lines indicated previously. This would then be followed by a discussion of each region in turn in terms of its particular characteristics. This was still the pattern in the postwar period, as in John Garland's (1955) *The North American Midwest*. Its last hurrah may have been Jean Mitchell's (1962) edited collection of regional studies of Great Britain.

A hallmark of this regional geography, therefore, was its highly descriptive character. This also applied to what few studies of particular regions there were. In J. W. Houston's (1959) study of the plain of Valencia in Spain, there is from the very start an attempt to provide some sense of the unity and distinctiveness of the region—to justify it as an object worth separating out from the mass of areal differentiation that is the world:

> The keynote of this region . . . has been the high degree of harmony between physical conditions and land use, together with a ready opportunism to change crop productivity according to the demand of external trade. This high degree of commercialism, however, has not removed the close intimacy which exists still between the peasant and his land. (1959, p. 166)

There is a strong man–land emphasis as in the reference to "the close intimacy which exists still between the peasant and his land." Sauerian influences are clearly in evidence, not least a strong sense of the visual as a means of apprehending the unity of the area:

> Scattered densely among the fields, like a city-in-the-country, are the white-washed cottages called "barracas" and the large, yellowed farm houses of "alquerias." Here and there the outline of the level plain is softened by the vertical clumps of palm, eucalyptus, or cypress trees. . . . Framing all this docile landscape are the mountains to the north and south, frail and brittle in the heat haze, and the more gently sloping hills to the east, stained red and yellow with the drought. (1959, p. 167)

The descriptive character of traditional regional geography, its concern with the unique or idiographic, which was supposedly (see Chapter 2) celebrated by Hartshorne and castigated by Schaefer,[18] along with its seemingly unscientific methods may have contributed to the crisis of geography as it developed in the 1950s: the view that owing to these emphases geography did not deserve a place in the university.[19] Against that backdrop one can begin to understand some of the enthusiasm with which many younger geographers, anxious about the future of the discipline, greeted the spatial–quantitative revolution and the promise it held for a more scientific geography concerned with explanation as well as description, something that is taken up in the chapter to follow.[20]

The Spatial Tradition

On the other hand, the way in which the spatial–quantitative work was to celebrate the spatial and explore with intensity its implications has led to a common belief that prior to it, it was little in evidence.[21] It did indeed usher in a self-conscious concern with the way in which human activities—industry, agriculture, cities—were organized over space and the spatial regularities that could be observed in that organization: clustering, regularities of spacing, areal specialization. Movements over space—migration, residential mobility, the diffusion of innovations—were subjected to the same framework of understanding; their spatial properties were what was to be explained, and space relations—the relative distances between adopters of an innovation, for example—were important in that explanation. However, quite aside from the fact that both the man–land and regional work were firmly within the spatial tradition, if drawing on a different concept of space,[22] the concept of the spatial at the center of the spatial–quantitative work had a long history. If anything, it enjoyed a prominence in the first 20 or so years of the 20th century that would only then be obscured by the dedication to the regional.

In an essay on what he has called "the invention of economic geography," Trevor Barnes (2000) indicated the way in which two foundational texts shared this same interest. The first was George Chisholm's *Handbook of Commercial Geography* (1889). Elsewhere Chisholm expounded, significantly, that

> It is the function of geography with respect to any class of phenomena that have a local distribution to explain that distribution in so far as it can be explained by variations connected with place in the operation of causes whose operation varies according to locality *or according to the relation of one locality to another.* (1908, pp. 568–569; my emphasis)

The other text, J. Russell Smith's *Industrial and Commercial Geography* (1913), which Barnes described as an American version of Chisholm's *Handbook*, is also notable. Underlining Chisholm's emphasis on "the relation of one locality to another," Smith outlined what he called a world economic geography of control and production. Control was located in northwestern Europe and the northeastern seaboard of the United States, since these were the areas that had capital to spare. The rest of the world was defined by its role as a producer. One can detect the same sort of sensibility in Bowman's *The New World*—a world in which some sort of spatial dialectic seems to be operating: As the opportunities for resolving the social problem within the United States through an expanding frontier evaporated, so attention would have to shift to an expansion of trade with the rest of the world (Bowman, 1928, Chap. 35).

Mackinder's work again is also exemplary. He is most famous for his explorations of spatial relations on a global scale with his Heartland theory and the contrasting spatialities of sea power and land power. But it is also clear in other of his writings, like his regional study *Britain and the British Seas*, published in 1907. Among other things, he talks there about:

- *Nodality*. Quite aside from the environmentalist emphasis on the role of natural waterways and the channeling of land routes by topography to create nodal points (p. 331), Mackinder recognized new nodalities that resulted from the convergence of railroads on new industrial centers: "It is obvious that modern industrial towns, based on local supplies of mechanical power or of metals, may grow large although lacking much nodality. . . . But if such communities endure they tend to create a kind of artificial nodality, as has notably happened with the great railway center of Birmingham. Even London-Westminster, twice made capital because naturally nodal in a high degree, has accumulated from its subsequent momentum a vast added nodality, as the focus of a radial system of paved roads and railways" (p. 330).
- *Spatial inertia*. "Should the significance of a town's nodality decrease, because, for instance, of mechanical inventions, or of new customs barriers, it does not necessarily follow that the town will forthwith degenerate. Much capital expenditure has been irrevocably fixed in it, or in connection with its trade, and great efforts may be put forth to improve its artificial nodality. Thus it may persist by *geographical inertia*, analogous to the mechanical inertia or momentum of a moving object. It is a 'going concern' with a goodwill based on the custom of trade, and is worth saving" (p. 330). This sounds awfully like the problem of local dependence, which was to be defined more generally some 70 or so years later.
- *The creation of new spatial divisions of labor*. Mackinder draws a contrast between an earlier urban geography in which towns served rural hinterlands and a later one that is being superimposed and in which the towns relate to each other through roles in a countrywide spatial division of labor: "At first a number of small market-towns . . . were scattered evenly over the more fertile parts of the country. They were local distributive centers at nodal points. . . . Now a certain number . . . are being selected for city-growth, while the rest dwindle with the general loss of rural population and the improvement of communication. . . . But it is characteristic of the rising places that . . . they obtain their renewed importance no longer as general distributors of the second or third grade, but by specialization of some definite type. . . . It follows that they are not self-sufficing after the manner of the old market-towns, but must supplement one another, or depend on some vast neighboring city" (pp. 337–338). Again, this was a long time before Doreen Massey talked about spatial divisions of labor.

This sort of emphasis was sharply attenuated with the shift of human geography's interest toward unique regions, but it did not disappear entirely. As early as 1933, Colby focused explicitly on the spatial character of urban form, arguing for it as the product of a balance between what he called centripetal and centrifugal forces and showing a sensitivity to the dynamics of urban land markets that would not resurface until the 1960s. Even earlier, Hartshorne (1927) recognized the crucial role of relative location in the explanation of industrial geography.[23]

Likewise, alongside the dominant interest in formal homogeneous regions, there persisted a minor countercurrent in which it is the nodal region, the region centered on towns and cities, that is emphasized. Mackinder had already advanced this idea in *Britain and the British Seas*, in his discussion of London and its wider hinterland. These arguments were then carried forward by British geographer Charles Fawcett and his student Robert Dickinson. Fawcett's work from the 1920s on was always informed by a strong spatial sensibility. In his 1917 article "The Natural Divisions of England," and despite the curious use of the word "nature" in the title, Fawcett anticipated by some years the more formal definition of the nodal or functional region.[24]

Dickinson, on the other hand, is a somewhat ambiguous figure. On the one hand, he seemed an exemplar of the geographers of the time. In the 1933 book *The Making of Geography*, coauthored with Osbert Howarth, Dickinson came out clearly on the side of the region and people–nature relations. The "essence of geography," he declared, "is the explanatory description of human occupance within composite natural regions" (p. 245). Yet, on the other hand, in some ways his work on urban spacing echoed Christaller's 1933 work on central places (Johnston, 2001). The conclusion of Dickinson's article on markets and market centers in England's East Anglia (1934) is significant:

> It will be evident from the foregoing study of the distribution and size of markets, that with the development of transport and modern organization, marketing activities show a tendency to concentrate in fewer centers. The distribution of mediaeval markets was such that all places were within two to four miles of one or more markets. In the early nineteenth century the larger market towns, at points of greater nodality, were located at ten to fifteen mile intervals. In recent years, there has been an increasing tendency to the further concentration of marketing activities in even fewer centers, now rendered easily accessible by both rail and road. (p. 182)

It is unlikely that Dickinson was aware of Christaller's work.

There is clear evidence, therefore, that the sorts of concepts of space that would become *au fait* in the course of the spatial–quantitative revolution were already circulating in the first half of the century, if as a minority presence. Even then, this is not to do entire justice to it. The concept of space that would be foregrounded by the spatial–quantitative revolution was of space as relative: One understood the locations of activities in terms of their locations relative to activities elsewhere—questions of accessibility and direction, in particular. And what was to be located could be reduced to a limited set of geometrical forms: points, as in the case of towns, or individual people, who might well be in motion as migrants, or indeed commodities to which the same qualification applied; or lines, like railway lines, the routes traveled by commercial airplanes, highway networks, and the like. Haggett set this out in his seminal text of 1965. What is curiously missing from his list are areas: an expression, that is, of the space-consuming properties of activities like factories, cities, housing developments, and, again, lines of communication for which land must typically be assembled and often through a process that is politically fraught. Hägerstrand drew attention to this in a 1973 paper. Relative space, he argued, was constituted not just by acts of spatial arrangement, distancing, clustering, and the like but also by the fact that it provided room, and activities needed room just as much as they needed to be in interaction with others at locations elsewhere.

The significance of this is the way in which space as room loomed so large in the early geopolitical writings of Mackinder and Bowman.[25] I shall have cause to take this up again later in the book; here I want to confine myself to a few remarks of justification. Both were taken up with what Hägerstrand would later describe as the matter of "providing room" but on a much grander scale than how he would express it through his time geography. Bowman was impressed by the closure of the American frontier and its implications for the need for American industry to, in effect, "find room" for its expanding flow of products in the rest of the world if the social contract between business and labor was to be preserved.[26] The idea of closure was also something that concerned Mackinder but less from the standpoint of a country for whom global hegemony lay in the future (the United States) than from that of a country (Britain) whose own hegemony was clearly being threatened: For him, the question was one of how to retain the room that had been acquired in the form of the British Empire.

THEORY AND METHOD

Field work provides us with the data, and, on occasions, takes us some way towards the elucidation of those data. It is an article of faith among us that field work is the essential basis of geographical study. When R. H. Tawney said that what economic historians needed was stouter boots, many of us paused to

consider the condition of our own shoe leather, and the cry
among us has quite properly been "field work and more field
work." To many, the field has been a welcome relief from the
methodological babble to which I am adding today.

Yet I suggest that the new cry might well be "Field work is
not enough." The map, to use F. W. Maitland's familiar phrase,
is a "marvelous palimpsest." Not all the ancient writing is legible
through what has been written since, but much of it is, and still
more of it is for those who have eyes to see. When, as geographers,
we gaze around, one question forces itself upon our attention;
it takes a variety of forms: "Why does this countryside look as it
does? What has given this landscape its present character?" The
moment we ask this question, that moment are we committed to
historical geography in one form or another.

—DARBY (1953, p. 9)

Prior to the spatial–quantitative revolution, and as Darby claims, it was
indeed "an article of faith" that field work was "the essential basis of
geographical study."[27] In this particular paragraph, he wants to use that
"article of faith" as a foil for his own view, based on his work as a historical
geographer, that is, that if we are interested in explanation we should also
draw on archival sources. But such a view had no effect on the field. With
some exceptions during the 1950s, like Roepke (1956), Farmer (1957), and
Smith (1955), whatever the methods deployed, the ends were largely of a
descriptive sort. And Darby's own work is no exception. His Domesday
geography of England was a massive undertaking, eventually running to
five volumes, but it was essentially about interpreting the source provided
by the Domesday Book and mapping the data. The study was devoid of
analysis and perhaps necessarily so given the severe limits of the archives
at that time. The Domesday Book of 1086 was indeed a remarkable compilation, but it is pure data and offers little in the way of clues as to how
they might be interpreted. But even for later periods when the archives
were altogether richer, they were put largely to descriptive ends.[28]

Parenthetically, and although he wants to make a point, we should
also note Darby's complacency about method: "To many, the field has
been a welcome relief from the methodological babble to which I am
adding today." However, it is a somewhat ambivalent complacency, since
he admits to adding to the "methodological babble" and given the time
at which he was writing he was clearly hedging his bets. The "methodological babble" was indeed a feature of the times and would eventually
be given more precise expression in the form of the spatial–quantitative
revolution. His admission is interesting for a further reason. When one
reads the article in question, it is clear that "methodology" for Darby was,
at least by that date, also about interpretation and the sort of interpretive

framework appropriate to historical geography. It is about how we should understand the relation between history and geography, time and space. In this regard, it stands as one of the few contributions to geographic theory that we have from human geography during the first half of the 20th century, along with those of a few others, like Mackinder and Sauer, even though Darby would have shrunk from such a word.

For to repeat: The emphasis in human geography was very largely descriptive. To read work from that period, whether in the form of professional papers or textbooks, is to be overwhelmed by maps, but maps that were rarely put to explanatory use. Rather, they provided a necessary complement to the narrative, which was almost entirely descriptive in character and typically focused on the relation to nature: the expansion of land under irrigation in some part of the world, the growth of hydroelectric power in another, the expansion of the settlement frontier elsewhere, the world distribution of coffee production, and so on.

Theory in a more implicit form, of course, as some framework of ideas could not be avoided. But it served less for purposes of interpretation and more for defining what was significant and what was not in a geographic study. Features of the world that related to the use of the land or the sea, the relation between geology and settlement patterns, or changes in land use patterns and plant disease, were selected. But as far as explanatory purposes were concerned, some sort of ecological "common sense" tended to step into the breach. Of course, villages were spaced along spring lines because people needed water; or, of course, they avoided the flood plain precisely because they did not like to be flooded out. We should not expect dairy cattle in tropical latitudes owing to the difficulty of keeping the milk fresh. In subarid climates, it made perfect sense to irrigate, and so on.

The central interpretive assumption was that if there were covariations across space they were between elements of the natural environment on the one hand and human activities on the other. The way in which regional accounts were structured is a nice expression of this. Almost certainly they would start with a discussion of what was called "the physical framework." This would then be followed, in turn, by sections on, successively, agriculture, mining, if there was any, then industry, and finally transport and towns. The material was arranged, in other words, to reflect an assumption about their decreasing dependence on relations to the "physical framework": It would start with agriculture and mining and conclude with those features less easily reducible to the facts of climate, geology, and topography and more a matter of what, in retrospect, we might attribute to spatial rather than to ecological logics.

From an explanatory standpoint, therefore, this was a human geography that was, for the most part and with some significant exceptions, impoverished. Given the rare acknowledgment of any sense of how social relations might structure the relation to nature, as in contemporary work on political ecology, we should not be surprised at that. To repeat: Theory, although it was never expressed as such, served as a criterion of what was significant. It was certainly not a matter of developing and evaluating alternative hypotheses with a view to the revision of theory. It was not a means of interrogating the relation between more abstract claims and the concrete and so adjusting those more abstract statements. Accordingly, people didn't *talk* about theory. There were no courses in university departments that could be called "theoretical" in their emphasis.

Perhaps if there had been a reliable means of testing alternative explanatory claims, things might have been different. Geographers did not have that. Map comparison could only take one so far. Relating the poleward boundary of cotton cultivation to a line defining 200 frost-free days or showing how building materials were a function of the underlying geology—at least premodern building materials—could provide some reassurance that map comparison worked. Most human geographies, however, were much more complex, and in those situations the possibilities of map comparison were rapidly exhausted.

Before moving on to how what limited map comparison there was came to be superseded, though, and in talking about theory and method, one last point needs to be made—one that is entirely in keeping with the lack of interest in theory: the remarkable lack of specialization of geographers that characterized the field in the first 50 or 60 years of the last century. This is especially apparent in the way in which many geographers practiced both human and physical geography. This was particularly the case with human geographers. Dudley Stamp is noted largely for his work in human geography, but he was also active in the interpretation of physical landscapes. A regional geography of Great Britain edited by J. B. Mitchell in 1962 is notable for the way in which the various contributors move between human and physical geography, particularly geomorphology and its relations to the underlying geology. American geographer Glen Trewartha made his mark largely in population geography, but he was also responsible for one of the best climatology texts for undergraduates in the late 1950s. Alfred Grove was a noted British human geographer, but he also did highly regarded work on desert landforms, as indeed did Yi-Fu Tuan early in his academic career and before he turned to more humanistic concerns. John Borchert was noted largely for his contributions to urban geography, but earlier in his career he had made respectable contributions to regional climatology.[29] In part this reflects the view that geography was about the relations between people and the so-called physical environment. It is also testimony to the enduring significance of field work as *the* method *par excellence*. This is the idea of geography as an intensely visual field of study (Driver, 2003). It was the way in which at one time generations of students were initiated into the "field": excursions into the countryside to examine the relation between geology and the form of the physical landscape, the relation between slope and land use perhaps, or variations in settlement pattern. Like so much else, with the spatial–quantitative revolution this was to become part of human geography's prehistory.[30] Geography was about to become much more fragmented, much more specialized than it had ever been before.

CONCLUDING COMMENTS

In retrospect, human geography in the first half of this century as practiced in the academy was an extremely conservative subdiscipline. There was an odd disinterest in modern, urban society, apart from a few visionaries like Fawcett and Dickinson. The notion of methodological or theoretical debate was utterly alien. There was no sense of forward movement. And lacking a strong sense of the social, human geography found itself closeted off from the other human sciences.[31] For the most part, complacency ruled. In Great Britain, this was absolutely the case. In the United States, though, there had been some rude shocks challenging that self-satisfaction and reflecting a view in some universities that geography was marginal to their intellectual purpose. The most notable of these was the closure of the department at Harvard, the most prestigious of American universities, in 1948.[32] For some this was a shock and resulted in a crisis of self-belief. But where could or would salvation lie? One answer would be the spatial–quantitative revolution, and indeed afterward human geography would never be the same. It is to that revolution that we turn in the next chapter.

NOTES

1. This is actually the date of the first appointment of a Lecturer in Geography at Cambridge. I have been unable to date the foundation of the School itself. See Stoddart (1989).
2. For an excellent discussion of Geography in the Ivy League universities, see Richard Wright and Natalie Koch's, "Geography in the Ivy League" (*www.dartmouth.edu/~geog/docs/ivy_geog.pdf*).
3. As in the titles of two of his more well-known books, *Civilization and Climate* (1915) and *Mainsprings of Civilization* (1945).
4. Likewise, "We must determine how much of our European and American energy, initiative, persistence and other qualities upon which we so pride ourselves is due to racial inheritance, and how much to residence under highly stimulating conditions of climate" (1915, p. 68).
5. "People who are subject to them (tropical diseases) cannot be highly competent. Their mental processes, as well as their physical activity, are dulled.

So long as a community is constantly afflicted with such disorders, it can scarcely rise high in the scale of civilization. Nothing is more hopeful for the tropics than the rapid progress in the control of these diseases. If they could be eliminated, not only might the white man live permanently where he can be only a sojourner, but the native races would probably be greatly benefited" (1915, pp. 60–61).

6. See also Darby (1951).

7. On the other hand, the work of historical geographer Andrew Clark, a former student of Sauer, is more akin to British historical geography of this genre. His book *The Invasion of New Zealand by People, Plants and Animals* (1949) is a good example of this.

8. "The Coal Measures of the concealed coalfields of Britain have existed for millions of years in the Phenomenal Environment but did not become geographically significant until geological discovery, improvements of mining techniques and demand for power brought them into the Behavioral Environment of British entrepreneurs" (p. 367).

9. This is not the sum total of the book. The writer was clearly aware of some of the fundamentals of the economics of air transport, but that makes the desperate and pervasive search for environmental relations all the more bizarre. At least in hindsight, this is the dominant impression left by the book. But at the time, and crucially, it did not appear so.

10. A second example is provided by O'Dell's (1956) *Railways and Geography*. This is very much more of the same. Of the seven maps included, one is entitled "The Influence of Geology on a Railway Profile" and a second "Crossing Mountains." Of the nine chapters, two are devoted to the relation to the physical environment: "The Land and the Rail" and the delightfully titled "Fog and Flood." A third chapter entitled "Motive Power" includes lengthy discussions of "gradients" and "curves" as well as a disquisitioning on the effects of the distribution of natural resources on the motive power used. So, "to countries lacking resources of suitable coal, electricity can obviate the need to import provided there are sites to generate hydro-electricity" (p. 78), while "adoption of diesel operation has been encouraged by their independence of water supply," so diesel traction is especially attractive where the availability of water supplies is in question (p. 81).

11. "The need of the world is not a new partition of territories between the Powers, according to this or that formula, but an ordered redistribution of population within the habitable regions of the world. The present congestion of certain lands and the emptiness of others does not correspond to the distribution of the world's potential resources. There are vast areas within the temperate and tropical zones whose climate, soil, and vegetation are favorable to human life, but which still life fallow. Their frontiers are closed to settlement for a variety of reasons, amongst them the desire for racial exclusiveness and the intention to protect certain economic standards against the intrusion of peoples of low material level" (p. 8).

12. An interesting exception was the work of C. Daryll Forde (referred to previously), particularly his book *Habitat, Economy, and Society* (1934), with the significant subtitle of *A Geographical Introduction to Ethnology*. I am grateful to one of the reviewers of an early draft of this chapter for reminding me of his significance. This work was published in 1934, underwent numerous editions, and was frequently a component of undergraduate courses in human

geography in British universities. I recall it being on a reading list from my own undergraduate days, my assiduous reading of it, and then my disappointment that there were no examination questions for which I could draw on my knowledge. This is one reason why I do not think it had much effect on the thinking of academic geographers of the time.

13. This would be in contrast to what would be called functional and nodal regions: regions which enjoyed a coherence through the spatial interactions common to them—a focus of flows on a central point or flows that defined areas as enjoying some common principle of spatial organization.

14. "Vidal de la Blache regretted what he could not help but observe. He considered that much that was best about life in France arose out of the range and balance of original communities to be found there. He considered, like many of his contemporaries, that the moral qualities of rural life were important to the nation and feared their decay" (Wrigley, 1965, p. 11). Lévy, the author of this particular entry (Lévy and Lussault, 2003, p. 985), added to this critique: "Enfin, si elle n'est pas sans parenté logique avec les idéologies du naturalisme progressiste qui oppose au réalisme des puissants l'utopie d'une nouvelle harmonie . . . cette géographie entre en phase avec ceux qui veulent conserver un équilibre menacé par la technique, les marchés et la ville" ["Finally, while it has a logical affiliation with the ideologies of that progressive naturalism which opposes to the realism of the powerful the utopia of a new harmony, this geography (i.e., Vidal's—KRC) is in phase with those who want to retain an equilibrium that is threatened by technology, the market and the town."]

15. On Fleure, see Pyrs Gruffudd (1994); and on E. Estyn Evans, see B. J. Graham (1994).

16. Bernard Marchand (2007) has pointed out its significance in France, where one of its products was a well-received book entitled *Paris et le désert français* [*Paris and the French desert*] (1947), by Jean-François Gravier. Gravier had been active on the extreme right of French politics before World War II. His view was that the increasing concentration of population in Paris was an important source of French demographic stagnation since the birthrate was significantly lower there. France's low rate of national increase had been a concern for successive French governments ever since the defeat in the Franco-Prussian War in 1871. Big cities were also a cause of loss of morality. The solution in Gravier's view was a radical decentralization from Paris in order to repopulate the "French desert." Marchand (2007) has written about a similar movement in Switzerland at the same time.

17. On the city and moral degeneration, see Pick (1989).

18. Though the degree to which Hartshorne actually stood for the idiographic is contestable.

19. For an outspoken example, see David (1958).

20. There were some notable exceptions to this rather uninspiring legacy. Spencer and Horvath, in 1963, addressed the question of origins of agricultural regions, anticipating by some decades the later interest in the social construction of space.

21. This is a belief that was sustained by Taaffe's authoritative Presidential Address of 1973.

22. Compare Gregory: "The production of geographical knowledge has always involved claims to know 'space' in particular ways" (2009, p. 707).

23. He also referred to the idea of optimal locations though in seeming inno-cence of Alfred Weber's work on the topic. Even so, there is evidence in the paper (p. 96) of the sorts of principle at the core of Weber's understanding: "The unit transportation costs are usually higher on finished goods than on raw materials, so that when the manufacturing process involves little or no loss in weight, and the raw material is non-perishable, locus with reference to markets is more important than that with reference to raw materials."

24. In designing his provinces, Fawcett had to come up with regional capitals. He clearly recognized the role of nodality. One of the best examples was Bir-mingham, which "is distinctly the commercial, financial, shopping, social, and intellectual focus of its region; it has a well-marked regional individual-ity, and is not, in matters of public opinion, in any way subordinate to any other center, a fact which is well illustrated by its Press and its public life. It is 'Town' for its region" (1917, p. 126). Along with others like Eva Taylor, Faw-cett was also instrumental in identifying a notable feature of the changing economic geography of the country: what was variably called the "coffin" or "axial belt" or relatively strong urban growth in an area with London at its Southeast corner and Manchester and Leeds at its northern apices.

25. On Mackinder, see Kearns (1984).

26. Interestingly, this is something that Neil Smith in his otherwise excellent (2003) book on Bowman does not discuss. This is surprising. He underlines the intellectual affiliations between Bowman and Frederick Jackson Turner while ignoring the latter's own arguments in favor of American imperialism subsequent to the closure of the frontier. See Stedman Jones (1972).

27. Compare Carl Sauer (1925): "Underlying what I am trying to say is the con-viction that geography is first of all knowledge gained by observation. . . . In other words, the principal training of the geographer should come, when-ever possible, by doing field work."

28. Darby's own study (1956) of the draining of the English Fens during the 16th, 17th, and 18th centuries, a large area of land subject to flooding from the sea, is a case in point. See also the book he edited in 1973.

29. It was also evident in classroom teaching. As an undergraduate, I recall a human geographer of very considerable repute, and deservedly so, teaching a course on the regional geomorphology of Wales.

30. In one respect this is to be regretted. This is because of the way in which field work brought physical and human geography together around the idea of landscape, as in the work of Sauer and Darby. There is a sense, though, in which the significance of this was never entirely grasped by geographers. There was certainly work by geographers that emphasized the relation between the visual and geography, as in classic texts of the 1950s like Stamp's (1946) *Britain's Structure and Scenery* or Gordon Manley's (1952) *Climate and the British Scene*, but some of the more influential texts, like Arthur True-man's (1949) *Geology and Scenery in England and Wales* or W. G. Hoskins's (1955) *The Making of the English Landscape*, came from outside geography; though that in turn may testify to the lay interest in geography as a study of landscape, since their influence was widespread: Both were published as Penguin Books.

31. Compare Taaffe, lamenting the effects of the areal studies tradition, which he termed "the integrative view" that had dominated human geography for much of the first half of the 20th century: "Another weakness of the

integrative view at this time lay in what should have been its greatest strength. The very thing it should have done most effectively, namely to bring geographers into closer contact with the other social scientists, it failed to do. In part this was due to the fact that geographers felt closer ties to geologists and historians; in part, to the fact that neither the methods nor the relatively few generalizations which emerged from geographic work formed an effective basis for communicating with the other social scientists" (1974, p. 6).

32. Yet as Smith (1987) has pointed out, closure of departments was not unreasonable given the inability of geography to present itself clearly. Accordingly, the committee appointed to look into its future at the university "was perplexed by its inability to extract a clear definition of the subject, to grasp the substance of geography, or to determine its boundaries with other disciplines" (p. 169).

CHAPTER 2

Long Live the Revolution!

It was not the numbers that were important, but a whole new way of looking at things geographic that can be summed up in Whitehead's definition of scientific thought, "To see what is general in what is particular and what is permanent in what is transitory." How shabby, parochial and unintelligent are the contrasting words of Sauer at the time, "The geographer is any competent amateur . . . we may leave most enumeration to census takers . . . to my mind we are concerned with processes that are largely non-recurrent and involve time spans beyond the short runs available to enumeration." But he was faced with a new generation, one that was both sick and ashamed of the bumbling amateurism and antiquarianism that had spent nearly half a century of opportunity in the university piling up a tip-heap of unstructured factual accounts. It was a generation that, without exception, was completely conventionally trained, but one that knew in its bones that there was something better, more challenging, more demanding of the human intellect than riffling through the factual middens and learning the proper genuflections from *The Nature of Geography*.
    —GOULD (1979, p. 140)

The spatial–quantitative revolution, taking off in the mid- to late 1950s and continuing through the 1960s was, I want to argue, the seminal event in the history of the discipline of human geography. It was never the same again and, for many, thankfully so. It sensitized a discipline to the importance of method and theory where there had been virtually no attention to either prior to that. Method, as we have seen, amounted for the most part to "field work," and there was precious little discussion in the literature as to what that might amount to in practice; it was as if it was to be exclusively absorbed through the pores of the skin in the process of actually doing it. "Theory," on the other hand, was a word that was virtually unknown in the vocabulary of the human geographer. The spatial–quantitative revolution changed that forever. Those resisting it might complain about how it was method for method's sake, which was true in a number of instances, but at least, and at last, people were talking about what appropriate method might look like, even if one might disagree with the sort of method under discussion. The same applied to theory. The spatial–quantitative revolution incorporated as part of its *modus operandi* abstract categories, which could then be filled with the more concrete. Peter Haggett (1965, Chap. 1) talked about the essential components of a spatial system: movement, networks, nodes, hierarchies, and surfaces, and finding new substantive domains to which to apply these ideas became commonplace. And once pattern had been identified—and there was certainly a lot of searching for it—there was the task of trying to explain, which was where theory came in.

One of the reasons it succeeded so well was that the times were ripe for it. This wasn't just a matter of the crisis that academic geography was experiencing at the time, and the notion that "science" might indeed be the right banner to march behind if it was to secure its position. The 1950s were also a very technocratic period. This reinforced the attractions of rigor and objectivity, however much these notions were to be critiqued in the future. And, of course, the problems of transportation and urbanization more generally that were surging to the fore in the rapid urban growth and suburbanization of the period found a ready object of interest for a human geography concerned with fashioning an expertise that it could claim as its own.[1] But that, in turn, imposed some limits on the spatial–quantitative revolution. Not least it meant that its application in human geography was very uneven. Urban, economic, and transportation geography dominated. Political geography was much less affected, and cultural and historical geographers were the ones left to snipe from the sidelines, albeit without much coherence until theoretical fashion swung to their advantage in the 1970s. It also meant commitment to a value-free version of social science. Accordingly, as the love affair with technocracy wore thin in the late 1960s, and as the civil rights movement suggested there were still disagreements over policy ends, so the spatial–quantitative revolution lost some of its allure and impetus.

While it lasted, though, it was hugely exciting for its participants. Compared with what had gone before in the 1950s, the field was indeed revolutionized. There was an enthusiasm, a fervor that is hard to convey to those who weren't there. The possibilities seemed endless. It was only a matter of time before a methodological and theoretical orthodoxy took over the whole field in the way in which neoclassical economics had triumphed elsewhere.[2] The spatial imagination in the hands of people like Bunge, Gould, and Tobler seemed to know no bounds. There were clear mavericks like Hägerstrand and Wolpert, but in fact all the leading lights were *sui generis*: Berry and central place theory, Tobler and new ways of describing space, Marble and transportation geography, Dacey and point patterns. Then, quite quickly, the enthusiasm lost some of its edge, and the spatial–quantitative work began to be looked at more critically. In addition, what made an important difference was that now the criticisms were coming from people who had actually practiced it and who, therefore, knew what they were talking about: people like David Harvey, Doreen Massey, Allen Scott, and Mike Webber. This was more convincing.

The Achilles heel of the spatial–quantitative work was its positivism, though recognition that this might be a problem came quite late.[3] What initially put the spotlight on it was the recognition that the spatial–quantitative work omitted questions of power, as indeed, given the way in which it drew for its theoretical inspiration on a location theory embedded in mainstream economics, it had to. This criticism made its first tentative appearance in the work of Wolpert and recognition that locations were not arrived at in a conflict-free manner but were, in fact, fought over. Harvey then deepened the critique and made it an external one by bringing a Marxist understanding of society to bear on human geography.

Since then, the spatial–quantitative work has tried to learn to live with alternative views of human geography that are much more compelling than the versions it contested in its heyday. Its universalistic pretensions have receded as the significance of context has been recognized, though it has found it harder to shed its empiricist and value-free skin. What is perhaps more disturbing is the rift that has emerged within human geography. The mainstream of the field, including the critical mainstream, is now occupied by people who have little time for the quantitative above the occasional tabulation or illustrative statistic. This is unfortunate. How to bring quantitative and qualitative methods into a more fruitful relationship remains a priority in the field.

SITUATING THE SPATIAL–QUANTITATIVE REVOLUTION

Nevertheless, the spatial–quantitative revolution was a crucial event and greatly raised the consciousness, at least of human geographers, of the fundamentality of the spatial tradition. Spatial organization or order, spatial interaction, and spatial pattern became central to the new vocabulary. Human geography excited because of its aesthetic qualities. There was order in the way in which things were arranged over the landscape. It was the human geographer's responsibility to unearth that order—since it wasn't always apparent—and then to show how it could be understood in terms of a limited number of theoretical postulates and structures. What happened was a sharp break with what had gone before—almost as sharp as one can imagine. It is with that rupture that this account begins.

In understanding the spatial–quantitative revolution, reference is quite commonly made to the Hartshorne–Schaefer debate about exceptionalism. Its significance is probably overdone. Schaefer provided a philosophical grounding for what the spatial–quantitative people were trying to accomplish, but they would probably have gone ahead anyway. Indeed, an interesting empirical question is the degree to which Schaefer's (1953) famous article was called upon merely to legitimate to the profession at large a transformation of method and theory that had already been carried out.

The starting point for the debate is Hartshorne's massive and often turgid tome *The Nature of Geography*, published in 1939, in which he set out to identify the essential nature of the discipline by examining its history and what he regarded as the authoritative statements coming out of that history. The outcome of this was thoroughly in accord with prevailing practice. The focus of geography, according to Hartshorne and following the lead of German geographer Hettner, was areal differentiation or chorology. Geography, Hartshorne claimed, was a science but not a law-seeking science. Rather, it interpreted *place-to-place* differences not through analytic strategies but through the synthesis or integration of diverse characteristics. This particular role for geography stemmed from its focus on space, in contrast to history with its focus on time and the other sciences, which had as their objects of study particular substantive domains like chemicals, the human mind, and society. In other words, and in hindsight, Hartshorne reduced space to place. Space, or more accurately differentiation over space, was seen in terms of unique combinations of attributes. It was this reduction that provided a point of critical entry for Schaefer.[4]

The response came in 1953 in an article entitled "Exceptionalism in Geography: A Methodological Examination." Schaefer's essential point was that geography was not exceptional in the sense of a subject matter that did not permit law seeking and explanations based on laws. All the sciences dealt with the unique: In psychology, no mind was exactly like another; in economics, buyers differed from each other, as did sellers; and so on. But this did not prevent the scientist from abstracting particular attributes and examining them for their association with yet other attributes. In geography, these attributes would be spatial so that the focus of geography was spatial patterns and the science of geography was the science concerned with the formulation of laws governing the spatial distribution of certain features on the surface of the earth. Geography, in other words, identified laws of location.

SPATIALIZING AND QUANTIFYING

The spatial–quantitative revolution is often considered as if its two distinctive elements, the spatializing and the quantifying, were a unity from the start: that to spatialize meant to quantify and vice versa. This is apparent in the claims of some of the later critics of something that they now call, pejoratively, "spatial science." What I want to suggest here is that this view needs revising: that what came together in the spatial–quantitative revolution were two relatively independent movements in human geography. There were clear moves toward the spatializing sometime before the spatial–quantitative work blossomed forth. Furthermore, spatializing without quantifying was to continue as an important aspect of the changes that occurred in human geography in the two decades or so after about 1950. The same goes for the use of quantitative methods. Initially, they were not used in a spatializing way. The concept of space as relative, which was to assume such significance, wasn't necessarily there. That was to continue to some degree as part of what I would call the spatial–quantitative revolution's soft underbelly: the use of multivariate statistics in attempting to "explain" some geographical variation but seemingly lifting the different observations out of their spatial contexts.

A Spatializing Moment

I should note here that Schaefer made no mention of quantitative methods. He talked about the formulation of what he called "morphological laws" but did not indicate any preferred approach to deriving them. The importance of this is that there was no necessary connection between the spatial and the quantitative methods that were to be brought together in the spatial–quantitative revolution. Rather, in the 1950s and prior to the concentrated engagement with quantitative methods as *the* approach for recognizing spatial regularity, breaking it down into its component parts, and analyzing it, there had been a quite substantial interest in the spatial as a framework of interpretation. This was true in both Great Britain and the United States.

In the United States the work of people like Edward Ullman and Chauncy Harris was exemplary. Ullman was an early student of spatial interaction and had a particular interest in commodity flows. He was also the first to introduce central place theory to an anglophone audience, though, interestingly, in the sociology rather than the geography literature (1941). Harris (1954) had taken the idea of population potential from the social physics literature and developed the concept of market potential as a way of understanding the distribution of manufacturing in the United States. Other than computing market potential for different locations across the United States, there was no engagement with quantitative methods; rather, from then on, demonstrating the relation between the geographies of market potential and industrial location, respectively, was a matter of map comparison, but quite convincing map comparison.[5] Furthermore, once the spatial–quantitative revolution got seriously underway, there were others who were much more spatializing than they were quantifying; an early example is Allen Pred. Brian Berry was Pred's graduate advisor but in both his thesis and his subsequent work, quantification was undoubtedly a minor theme and often scarcely evident at all. And Pred was certainly a spatializer of considerable significance. Another case was William Bunge.

In the introduction to his *Locational Analysis in Human Geography*, which appeared in 1965, Peter Haggett fondly recalls his cohort of undergraduate geographers at Cambridge in the mid-1950s "that included men as talented as Michael Chisholm, Peter Hall, Gerald Manners and Ken Warren." These all made their contributions to an emergent and spatializing "new" geography. Chisholm is noted for his book *Rural Settlement and Land Use*, published in 1962. Much of the focus there is spatial regularity, and the quantification is fairly elementary: the use of means for concentric bands defining yields per acre, for example. Interestingly, however, the simple nature of the methods used did not detract from the defensibility of the conclusions arrived at. Hall and Manners, on the other hand, were in part the legatees of a spatializing tendency that went back to the debates of the 1930s about location and planning in Great Britain and, in fact, the work of Fawcett on the changing location of population in the country and what it said about its shifting economic geography. Peter Hall was later to be important in the establishment of a new journal, *Regional Studies*. This had to do primarily with policy issues of the growth of large towns, what to do about getting industry into the depressed areas, and much of the work was conceptual and theory oriented rather than technical in the narrow sense.

This is far from claiming that the only precursors of the new advocates of a spatially focused geography in Great Britain were associated with Cambridge. One name above all that should be mentioned is Wilfred Smith, who was at the University of Liverpool. Smith (1955) was writing about Alfred Weber and testing his theses regarding the significance of weight loss and weight gain in the manufacturing process for the localization of industry some time before the spatial–quantitative revolution would discover location theory. Another person of note is Michael Wise. He conducted pioneering studies of industrial districts at least 30 years before Allen Scott would stamp his name on the topic (Wise, 1949). His careful empirical studies of the gun and jewelry quarters in the English city of Birmingham are still worth revisiting, and his theorization recalls some of Scott's own arguments about the agglomerative tendencies of vertically disintegrated labor processes.[6]

There was, therefore, an emerging focus for human geography and an emerging vocabulary. People started thinking in terms of spatial regularity, as in Chisholm's testing of Von Thünen's theory of agricultural zonation or in Ullman's theory of spatial interaction and the role of distance (1956). Later there would be talk of "spatial organization." The sense of some spatial predictability began to seize the imagination. People started looking for regularity—and they found it. The spatialization that this gave rise to was to become a constant theme in human geography in the second half of the 20th century,[7] even while concepts of space became significantly enriched and the emphasis on quantitative methods in evaluating concepts and theorizations of space waned. But early on, the application of the spatial imagination to new subject matter could be quite stunning.

A notable instance was the 1962 work of William Bunge: *Theoretical Geography*, a book of quite stunning vision. Bunge showed a remarkable facility for taking a social science concept and, through spatializing it, not only enriching it but at the same time underlining what human geography shared with the other social sciences. Regionalization for him became the social science practice of classification applied to locations. Instead of, for instance, classifying people and placing them in social classes, one placed locations in regions. The criterion for classification was the same in the two cases: Minimize variation within classes and maximize it between.

A second rich conceptual vein that he exploited was work on time series. Economists extrapolated these in order to come up with predictions, so why not extrapolate geographically? In other words, one could take a fragment of a wider geography of settlement and lines of transportation and predict spatial arrangement beyond its margins, drawing, of course, on theory regarding the distances between towns of different size, geographic trends in population density across the area, and so on (see Figure 8.22, p. 247, in Bunge, 1964). Some years later the idea of spatial prediction was to be applied in striking fashion by Waldo Tobler (Tobler and Wineburg, 1971). This was his so-called Cappadocian speculation, in which he tried to locate a "lost" city in Anatolia by applying a gravity model to study of the geographic distribution of coins minted there and located in various archaeological digs.

A Quantifying Moment

Obviously, a defining characteristic of the spatial–quantitative revolution was the adoption and spread of so-called quantitative methods. But just as you could spatialize without quantification any more serious than the use of averages for distance bands, plotting measures of newspaper circulation around towns (Odell, 1957), or mapping the frequency of bus services (Green, 1950), so you could quantify without engaging with ideas of spatial organization.

The hallmark of the spatial–quantitative revolution would be its embrace of a particular concept of space: that of space as relative. One was to understand what happened at particular locations in terms of how they were geographically situated with respect to other locations: notably their accessibility as evaluated through measures of distance, direction, and connectivity through various media of transportation.[8] This was a concept of space that had led, by and large, a subterranean existence in human geography during the first half of the 20th century. Then, and in accord with the dominant focus on regions and areal differentiation, the foreground had been taken up by a quite different concept of space. This was space as absolute. Locations or places were located not with respect to each other, as in concepts of space as relative, but with respect to something devoid of the material: something empty called space. This was identifiable through some arbitrary grid or system that significantly made each location unique and not, as Schaefer recommended, comparable with other locations—as in "distance from Chicago." A location might, therefore, be at a particular intersection of longitude or latitude; or it might correspond to a part of a system of administrative subdivisions—a state, a county, or a municipality. One could describe data as "geographic," therefore, but with limits as to what "geographic" entailed.

This was to be the case with much of the early quantifying work in human geography, which involved simply the application of conventional descriptive statistics to the following sorts of geographical data: data for counties on such variables as voting for a particular party, population density, precipitation, wheat yields, employment structure, and so on. Early quantification gave limited attention to what would later be known as "location variables." This is very clear in the early work of the so-called Iowa School, where quantification was formalized in terms of an iterative procedure that included the use of multiple regression to test hypotheses, and then the examination of residuals from regression in order to identify other "variables" that might be important in "explaining"[9] a particular geographic distribution. These new variables might be included in a reformulation of the regression model in order to achieve a higher level of "explanation." People strongly associated with this approach were Harold McCarty (McCarty, Hook, and Knos, 1956) and Edwin Thomas (1960). They were by no means the only ones, though. Arthur Robinson at Wisconsin did very similar work,[10] as indeed did British geographer Peter Haggett (1964).

THE SPATIAL–QUANTITATIVE REVOLUTION
"PROPER"

In short, the spatializing and the quantifying impulses could proceed largely independent of one another, and this was to continue long after the spatial–quantitative revolution "proper" got underway from the late 1950s on. This was the combination of the two. Spatial organization or regularity, drawing on concepts of relative space, would be the substantive focus. Quantitative methods would then be the preferred method of identifying what regularity there was and then analyzing or, as some would eventually have it, "modeling" it.

Spatial Organization

The legacy in terms of what we now know about spatial organization is an impressive one. Among other things, and simply by way of illustration, we should acknowledge the following:

1. *Work on urbanization and urban spacing.* Drawing on the pioneering work of German geographer Walter Christaller on central places (1933), a large and impressive literature was put together demonstrating the clear regularities that exist between, *inter alia*: the various service functions that different cities perform; their different sizes; their spacing relative to cities of similar size (smaller cities closer together, larger cities farther apart); and travel-to-shop distances. Much of this was Brian Berry's work or the work of his students (e.g., Berry 1964, 1967; Berry, Barnum, and Tennant, 1962). Of course, there are distorting factors like the presence of industry or changes in population density. However, once these are taken into account, the regularities remain impressive ones.

2. *Work on the geometry of transportation networks.* One of the more notable contributions here was that of Christian Werner (1968) on what he called the delta-wye transformation. This refers to changes in the geometry of transport networks—railroad, highway, or airline—as the populations being served change. Initially, three cities at the apices of a triangle may be served through a fourth, larger city that is central to them all: Passengers traveling from one apex city to another have to go "a long way round"; therefore, passengers traveling from one apex city to another are "bundled" with those simply traveling from the fourth, central city to one of the apex cities. This bundling is due to an adverse ratio of fixed to running costs. Constructing a railroad between two apex cities cannot be justified in terms of the relation between the fixed investment and the amount of traffic to be expected. But as the populations of the apex cities

increase, so potential traffic increases, the economics changes, and the "wye" form gives way to a "delta" one.

3. *The gravity model.* The inspiration here is Newtonian physics and the claim that the attraction of one body to another is directly proportional to their masses and inversely proportional to the distance separating them. Although there is no clear theory as to why it should apply to very different forms of spatial interaction, it does: migration, trade, shopping trips, marriage patterns, where the students at a particular university come from, even the frequency with which news stories about particular places are included in a particular newspaper. It, therefore, provides a benchmark for the examination of seeming exceptions.[11]

Taking the Measure of Space

A focus on spatial organization required appropriate methods. The original focus on simply applying regression to geographic data quickly gave way to a more concentrated search in other literatures for techniques and methods that might have been devised more specifically for the sorts of data and problems geographers were interested in. This was especially clear in the case of the people at the University of Washington, including William Garrison and his graduate students (Dacey, Berry, Marble, Tobler, and Nystuen in particular). Some examples include:

1. *Borrowing methods from quantitative plant ecology for the analysis of point patterns as in quadrat count analyses and nearest neighbor techniques.* A common field work technique of botanists was to divide study areas into squares or quadrats. Within each quadrat, counts were made of the frequency of occurrence of particular species. The resultant frequency distribution of quadrats with 10, nine, eight, and so on instances could then be used in the development of measures of the degree of clustering of those instances. This technique found application in the work of geographers on settlement patterns (Dacey, 1968) and on the distribution of adopters of particular innovations (Harvey, 1966): Did innovators tend to cluster, or was their distribution random, for example?

2. *Borrowing methods from operations research for the identification of normative geographic patterns with which the actual could be compared.* Garrison was especially interested in this, as is apparent in the sequence of three review papers he wrote (1959-60) for the *Annals of the Association of American Geographers.* An example is the use of linear programming solutions (allocation of resources under constraints of demand and supply) to the so-called transportation problem: Given a set of demands for a product at a set of locations and a set of supplies at a, not necessarily exclusive, set of locations, what should be the allocation of supplies to demands in order to minimize total transportation costs? The normative pattern

established can then be compared with what has actually transpired or provided to a client as something to consider adopting.[12]

3. *Borrowing methods from mathematics in the form of graph theory for the description of networks.* Again, Garrison (1960) was to the fore, to be joined later by Marble (Garrison and Marble, 1965). A major outcome of this was recognition of the idea of the connectivity of networks and of the connectivity of particular nodes in networks: highway, airline, and railroad.

There was also an interest in the modification of aspatial statistics and techniques so that they could be applied to spatial problems. In an early paper, Tobler (1966) tried to adapt techniques for the description of time series (e.g., moving means) to what he called spatial series. A couple of the measures he came up with were (1) the spatial moving mean for smoothing out geographic variation (this could be applied at a variety of geographic scales and so afforded an entrée into the problem of scale) and (2) a spatially moving variance—to provide a measure of variation in variability! Another instance of the adaptation of aspatial methods to geography is the idea of trend surface analysis. The idea here was to summarize trends in geographic distributions through fitting a multiple regression model: specifically, regressing some geographically varying characteristic like population density on geographical coordinates (eastings and northings). This could provide estimates of the fit of linear trends or, by the addition of powered terms, nonlinear trends.[13]

On the other hand, and in these early years, there was little attention to some of the distinct problems arising when describing geographically organized data through quantitative methods. The problems of analyzing areal data—data for counties, for example—which emerged with the early work at Iowa and Wisconsin were examined first by sociologists (Duncan, Cuzzort, and Duncan, 1961). Even then, it was only later, in the early 1970s, that the quite crucial problem of spatial autocorrelation was taken up and also that there were real attempts to come to terms with problems of spatial aggregation, as in the work of Stan Openshaw (1984). Work on the gravity model underwent a similar evolution. It was imported from outside geography, and it would take some time before the problems that a nonisotropic plain posed for evaluating the so-called effect of distance came under critical scrutiny with the pioneering work of Curry (1972).[14]

MODELS

Aside from the mechanics of quantification, another major theme in the early work that should be acknowledged was that of the model. Reference to models was widespread. An early and very influential compendium vigorously announced models as a theme in its title (*Models in Geography* edited by Chorley and Haggett, 1967). Models were defined as simplified and/or idealized representations of the real world and could come in many different forms.[15] In statistical and mathematical models, reality was represented by various symbols and operations, which could be compared with the real world and also manipulated to resolve "What-if?" questions. Monte Carlo simulation models were one early form of statistical model deployed to great effect by Swedish geographer Hägerstrand (1957, 1965). Other forms included scale models, as in maps, or analog models, in which a model for one substantive domain was applied to some other. Swedish geographer Lövgren applied interregional input–output analysis, originally developed for understanding commodity flows to migration (1957). The gravity model was applied to an increasingly wider range of substantive domains, including the geographic origins of news stories in newspapers to the location of lost prehistoric cities, as in the work of Tobler and Wineburg (1971) noted previously. Models were to be compared with the real world, they could be manipulated in order to afford insight into the real world, and they could be subject to correction in an iterative fashion.

A fundamental distinction was, and remains, that between theoretical and empirical models. In theoretical models empirical expectations are *deduced* on the basis of a closed system and assumptions about the actors: preferences, budget constraints, technologies available, and so on. These are common in economics, and during the spatial–quantitative revolution urban and economic geography tended to model themselves on that field. Examples include Christaller's central place theory: Assumptions about the distribution of population, shopping trip behavior, threshold markets for different retail goods or services, and distances consumers are willing to travel in order to purchase a good or service generate a pattern of central places of different sizes. In this pattern, larger central places are separated by longer distances than smaller central places, shopping trip distances are related to the size of central place that is the destination, and so on. In other words, in theoretical models, assumptions are made about spatial behavior, the context in which the behavior is unfolding, and then deductions are made about resultant spatial patterns. They are typically tested using regression, as in Berry's pioneering work referred to previously. A weakness, however, is that the coefficients resulting from the regression tend to vary over space and time, suggesting that the deductive model provided an incomplete specification of all the relationships necessary to explaining a particular pattern, which is not surprising given that a closed system is assumed in deductive models—invariant contextual conditions, highly simplified ideas about human action—and in the real world things are different.

Far less ambitious are empirical models, which are typically based on induced relations. These lack the formal character of deductive models. They typically incorporate sets of relations that have been empirically observed elsewhere or at least loosely hypothesized, perhaps through a process of casual empiricism. A common form is the multiple regression model, which postulates relations between a dependent variable—what is to be "explained"—and a set of independent variables. The latter are chosen on the basis of a reading of the literature on the relevant topic, on what others have done, and on lacunae in their models: variables that were omitted but for which there are good reasons for inclusion. Alternatively, empirical models could draw on means and variances to establish or confirm some generalizations about magnitudes and the extent of variation: for example, interregional convergence of per capita incomes; evaluations of settlement dispersions—clustered, random or uniform; the use of segregation and localization indices.

It is important to ask just why the concept of model became such a crucial part of the lexicon of the new spatial–quantitative geography in the 1960s and has indeed remained with us. In part, it was an attempt to assert the scientificity of the new approaches: Modeling was part of the language in the prestigious natural sciences and also in the most prestigious of the social sciences: economics. But there is another answer and that has to do with the problem of theory. Quantification provided a method in a very narrow sense. It provided, in particular, a means of testing hypotheses, as in regression or tests of statistical inference. But where were the hypotheses to come from? And how were the results to be incorporated into a growing body of theory? The crucial medium here was the model. This was because through its idealizations it provided a set of expectations about the so-called real world, which could be evaluated and adjusted in cumulative fashion. As geographers moved on from an early fascination with methods for their own sake to actually applying them, therefore, the concept of model satisfied an essential need.

The Question of Theory

Even so, if one were to model, there had to be theoretical postulates. In fact, the spatial–quantitative geographers talked a lot about theory. Hypotheses were tested, theoretical claims were made, and papers typically ended with a discussion of "theoretical implications." On the other hand, they didn't think much about it in a critical sense. In deriving some theoretical base for their work, there seems to have been two major tendencies. The first was to derive theory from what were essentially empirical generalizations, inferring abstract notions of individual spatial behavior that could, at best, be reconciled if one had the inclination to the utility maximization assumptions of mainstream economics. The classic instance here is the gravity model, which lacked any theoretical basis other than the Newtonian analogy but which was interpreted in terms of least effort behavior. Some of the work on point patterns dealt with very simplistic notions of competition over space to account, on the one hand, for tendencies to uniformity (avoiding competitors) and, on the other, for tendencies to clustering understood, if at all, by appeal to Hotelling's location problem of ice cream sellers on a beach. This sort of reasoning could also issue in broad universals—very broad!—regarding space relations. For Bunge, geography's central problem was what he called "the nearness problem" (p. 211): "finding the spatial arrangement of objects . . . and placing these objects as near to each other on the earth's surface as possible" (p. 211). In this formulation, spatial processes consisted of movements while spatial structures were the resulting arrangement of things across the earth's surface.

This understanding of theory, as something, in effect, derived from empirical regularity,[16] clearly lacked any understanding of the social structuring of Bunge's location problematic. Ideas about the social would make their way into human geography in the form of location theory. This was an outgrowth of mainstream economics or, as in the case of Lösch, what amounted to an internal critique of it. Christaller's central place theory was the starting point for Brian Berry's (1967) extraordinarily fruitful investigations into the distribution of urban centers of various sizes. Earlier work by Michael Chisholm (1962) on agricultural land use drew heavily on Von Thünen, and Alfred Weber featured prominently in the work of those seeking to decipher industrial location patterns. Applications of Von Thünen's theoretical structure by urban economists like Alonso (1960) and Muth (1962) to the case of urban rent then allowed inferences to the distribution of population in cities, which were also of interest to the spatial–quantitative geographers.

But just as the new quantitative geography was slow in exploring the difference that space made for quantitative analysis, as in the question of spatial autocorrelation, the same seems to have happened with location theory. The fact of spatial difference, of locational variation, is a massive challenge to the basic assumptions of location theory, but there was some delay in recognizing this.

For a start, space plays havoc with the assumption, imported into location theory, of perfect competition. Bringing space into the analysis has to introduce a monopolistic element into competition that is unwelcome from the standpoint of much of traditional location theory. Competition over space has to be monopolistic, as the owner of any mom-and-pop store will affirm, since otherwise they would be overwhelmed by the more distant superstores. In other words, location can provide some insulation from competitors; or through, for example, agglomeration effects, it can provide production cost advantages that allow firms elsewhere not enjoying such advantages to be swept into oblivion and facilitate the emergence of a spatial monopoly in the particular product at issue.

Location theory postulated a state of spatial equilibrium, which, again, was contradicted by the difference that space makes. Spatial equilibrium turns out to be a pipe dream. Through spatial externality effects, any change in location sets in motion changes elsewhere, and the fact of investment in fixed capital of long life—notably buildings but also highways—makes adjustment to the sort of equilibrium imagined by Alonso (1960) and Muth (1962) so slow as to virtually preclude their very possibility. Although there might be a tendency in that direction, there will be new disturbing elements with entirely new spatial externality effects that will push the system away from that tendency (Roweis and Scott, 1978). The fact of the spatial immobility of land and the improvements to it would then preclude a return to equilibrium before new disturbing events occurred.

Applications

The spatial–quantitative revolution enjoyed a close association with applied work. This, of course, has continued, with the advent of geographic information systems (GIS), which have many and varied applications in both business and the public sector. In the early days, though, urban and regional planning was an important stimulus. Some of this worked through the Regional Science Association (RSA). Regional science was the brainchild of Walter Isard, an urban and regional economist who proved himself an able missionary. As Trevor Barnes points out in his (2009) entry in *The Dictionary of Human Geography* published by Blackwell, regional science is very difficult to define but it shared considerable overlaps with the work of spatial–quantitative geographers. Human geographers were enthusiastic attendees of the annual meetings of the RSA in the 1950s and 1960s, but so too were planners and urban economists, including people like William Alonso, Benjamin Stevens, Charles Leven, Richard Muth, John Kain, John Friedmann, and Melvin Webber. Perusal of *Papers and Proceedings of the Regional Science Association* or *Journal of Regional Science* from those days indicates just how significant the input of planners and urban economists was, along with that of human geographers.

We should also note that the impact of the spatial–quantitative revolution on human geography was very, very uneven. The most significant impacts were in economic, transportation, and urban geography. It was in these fields that the idea of location theory was most applicable. Some sense of this applicability is apparent in Smith, Taaffe, and King's edited work *Readings in Economic Geography*, which appeared in 1968 (and a very good book it is too). Virtually all the names of those prominent in the spatial–quantitative revolution were associated with one of these areas. Berry was an economic and urban geographer; Dick Morrill's work was largely in urban geography; Garrison, Marble, and Taaffe made their reputations in transportation geography.

However, the fields of cultural, historical, and political geography were much less affected. In fact, it was cultural and historical geography that provided the core of resistance to the spatial–quantitative work. Their complaints were ad hoc and lacked a strong theoretical basis, but that would change in the 1970s with the advent of humanistic geography. Political geography was slightly different in that the area of voting studies—hitherto known as electoral geography—readily lent itself to quantification. This was a minor growth area in human geography in the 1960s and 1970s (Taylor and Johnston, 1979). What is also interesting is the way in which this attempt at rapprochement with the spatial–quantitative led to a conscious attempt to spatialize the field, something that was to lead to increased interaction with political scientists because now it seemed that geographers had something distinctive to say about voting. Even so, this was an exception. For the most part, political geography remained apart and the reason was not difficult to see: The spatial–quantitative revolution had very little time for the question of power, an issue discussed at greater length in Chapter 3.

And the "Old" Geography

For the most part, the human geography of the first half of the 20th century seemed to have been entirely superseded. The old concern with people–nature relations had passed almost completely out of the picture as the focus shifted to the urban, to transportation, and to industrial location. Regional geography, as far as spatial–quantitative geographers were concerned, was dead, although the region itself maintained a continuing presence. Regionalization was, as Bunge had noted, a matter of classifying locations and their substantive attributes into classes such that the variance within classes would be minimized relative to that between. This then became the basis for the development of various algorithms designed to accomplish just that. Regional identification could be precise. This was regarded as a major advance over the subjective fumblings of what many regarded as human geography's Neanderthal period. Just why human geographers should continue to be interested in regions raised interesting questions that would eventually help to challenge their theoretical assumptions. In part, it was that any geographical research was itself located. Region would enter in as researchers struggled in their modeling to hold off as many complicating factors as possible in order to reveal the spatial regularities that, it was believed, were surely there. But that meant drawing upon some ideas of regional order. Coefficients would then show some variation between one region and another, although not always in self-explanatory ways so that region became something of a black box. So while the spatial–quantitative geographers devalued the significance of the region, it had an uncomfortable obstinacy about it.

CRITIQUE AND REFORMULATION

The reception of spatial–quantitative work in human geography at large was anything but unproblematic. It generated huge amounts of controversy and opposition; all the standard bearers of what they were calling "the new geography" had tales to tell of the difficulties of getting their work published. On the one hand, quantitative methods were being touted as the medium for making geography "scientific." The lack of ambiguity, the precision inherent in quantification, made it an ideal vehicle for re-creating geography in the image of a law-seeking science embarked on building knowledge of a cumulative kind. These were things that the geographer's traditional tool—map analysis and comparison—could not possibly do (McCarty and Salisbury, 1961). But there was an avalanche of ripostes and skepticism from more traditional voices in the discipline. It was suggested, for example, that:

1. The findings of the "new geography" often amounted to little more than statements of the obvious: that of course, for example, land values would be correlated with increasing distance from the central business district (CBD) or that migration would drop off with distance from its point of origin. In a widely noted, even notorious, comment by Dudley Stamp, at that time still the doyen of British geographers, on an early paper of Peter Haggett, the observation was made that the use of quantitative methods seemed to be a case of using a sledgehammer to crack a nut (Chorley, 1995, p. 359).

2. Methods were being put ahead of theory and findings of real substance; there was too much research on method for method's sake.

3. The real point of research was to explain, not to generalize; a generalization in the form of a correlation or regression coefficient, for example, was only specifying what had to be explained, and so the new geography was a very partial answer to achieving the real aim of any science, including human geography.

Much of this debate was markedly intergenerational in character: a case of young gladiators against an old guard resisting change.[17] Some of the points raised also had merit even if overgeneralized. But not all the findings were cases of stating the obvious, by any means, and even when they were they were being placed in new relations to yet other findings in more comprehensive and coherent theoretical structures. Moreover, the old guard was at a disadvantage. This is because it had little to offer other than a return to approaches to the field that stood discredited. There could be no intellectual advance using them, as I pointed out at the end of Chapter 1, and by emphasizing the language of science and cumulative discovery, the advocates of the spatial–quantitative had successfully claimed the high ground. Drawing on the experience of other social sciences, they could credibly claim to be on the right side of history and, given the technocratic tenor of the times, they could get away with it. This is not to say that serious intellectual challenges and issues were lacking. Rather, spatial–quantitative work developed. It did this through the unearthing of contradictions. These might be within its own intellectual foundations or emerging from the accumulating body of theoretical claims. So there was some reformulation, which in turn added to the sense of a framework of understanding capable of renewing itself, a framework that was inherently dynamic, therefore.

Spatial Separatism

An early issue is what is known as spatial separatism. The focus of critique here is the assumption that a spatial domain could be identified independently of any substance. This is very clear in Bunge's (1962) view in *Theoretical Geography* that the appropriate language for the new geography was geometry. The most argued-out critique of this came from Sack (1972). To exemplify his point: To argue that a city expands in area because the distance of its boundary from its center increases does not constitute an explanation, even though it expresses a geometrical law. Likewise, "distance" tells us nothing that is substantively useful unless we know something about the time it takes to traverse such a distance, how costly it is, and so on. So the "friction of distance" that spatial–quantitative geographers of that time had a tendency to refer to was a misnomer: It is substances that create the friction. In other words, if geographers wanted to explain substantive patterns like the spacing of cities, the location of industries, or flows of commodities, then questions of space and spatial relations could not be separated from the substance—people, material artifacts like highways—to which they referred.

An important consequence of this analytical indissolubility of space and matter was the recognition that all social sciences could be seen as concerned with spatial relations. Geography, therefore, should not claim an independent status on the basis of the spatial variable. Parenthetically, we should note how the belief that it could parallels Hartshorne's earlier argument about the exceptionalist nature of geography. According to Hartshorne, geography was exceptional because, owing to its interest in specific places, it could not be a generalizing science. In the revised spatial version of the exceptionalist doctrine, geography's enduring interest in distribution again asserts itself but this time as space or, more accurately, *relative* space. And while, according to the advocates of the spatial–quantitative revolution, the focus on relative space allowed geography to become a generalizing science, it was also claimed that, in addition, it gave geography a role independent from that of the other, more substance-oriented social sciences. Spatial separatism, in other words, and as it was so denoted by Sack (1972), was just another form of the exceptionalist argument. Incorporating substance theories from the social sciences did something to blunt Sack's criticism, though not entirely. It also raised new issues of a critical nature. This had to do largely with the nature of the substance theories used. These initial explorations continue to the present day, though along with an accumulating body of criticism.

Behavioral Geography

One of the sources of substance theory turned out to be psychology. The use of neoclassical economics in understanding geographies spawned criticisms of its assumptions, including those of rational behavior under conditions of perfect information. This led to an interest in, for example, learning theory; the effects of imperfect information on decision making; and the replacement of assumptions of rational with satisficing behavior. This was the so-called behavioral revolution in geography that was at its height in the late 1960s to early 1970s.

From the standpoint of more inductive approaches, behavioral geography was also a response to the fact that the models did not always fit the data. As Hägerstrand (1957) noted in his work on migration, there were deviations from the distance decay rule: Migration between some points of origin and destination would be wildly underestimated by the gravity model. There was also a creeping suspicion that the space in which locators operated might not be the same as the supposedly objective one to which the experts—the geographers—had access. This led to two significant and related, even overlapping, ideas: that of perceived space and that of the variable information that people had about different locations.

The first of these set the scene for the idea of the mental map. This was initially normative in content, and Peter Gould was the initiator of this line of research.[18] Gould was interested in people's preferences for places as places in which to live. No matter where respondents were currently located, their preferences tended to be quite uniform (Figure 2.1). So in Great Britain, areas like the south coast emerged as more desirable than

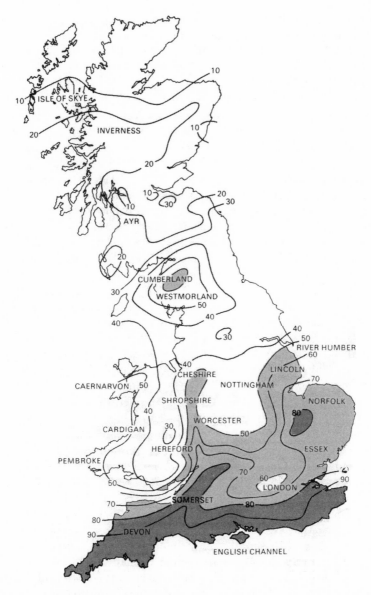

FIGURE 2.1. Mental Maps: The view of Great Britain. This is the rankings of counties school leavers would prefer to live, if they were free to choose—based on averages. The school leavers were from 23 schools scattered across England, Scotland, and Wales. From Gould and White (1974), p. 82. Copyright 1974 by Penguin Books. Reproduced by permission of Taylor & Francis Books, UK.

ones like, for example, Scotland. But there was also a local component in which people expressed strong preferences for their current location. This idea of place preferences could be extended by reference to what the data indicated as to people's choices. A notable argument here was that of Rushton on what he called space preferences (1969). This was an attempt to spatialize the preference schedules of neoclassical economics by examining not actual consumption behavior but actual travel behavior. In this way, Rushton was able to challenge the assumption Christaller had made in central place theory: that people patronized the nearest town (or central place) offering the goods or services they wanted. The problem with Rushton's approach, however, was its circularity: deriving preference schedules from actual behavior in order to predict that same behavior.

To this idea of preferred places or locations was later added a more cognitive approach to mental maps. If seeking to determine, for example, what sort of map of major urban centers in Ohio people would provide, researchers could give subjects a map of Ohio and ask them to insert the locations of the 9 or 10 largest urban centers, and the maps could then be analyzed for inaccuracies or biases. The sort of conclusion coming out of this research was that people tended to underestimate further distances and overestimate nearer distances to their current locations. A variant on this sort of research had more to do with people's understandings of regional classifications. Everybody has an idea of the South, the Midwest, and so on, but what exactly are those understandings (Cox and Zannaras, 1973)? One important feature of this mental map work was how it became an area of research in its own right and connections with location theory became quite loose, even though the deviation of results from models had been one of the stimuli to the exploration of mental maps.

In contrast, a second area of emphasis tended to have closer ties to researching actual locational outcomes. This looked at variations in the information people had about places: not just relative locations but the characteristics of different places. An early and influential example of this was Wolpert's (1964) study of land use in part of southern Sweden. He was particularly interested in labor productivity and in comparing optimum values with actual values, calculating the former through the use of linear programming. The comparison revealed quite substantial deviations, which Wolpert attributed in part to the notion of satisficing decision making on the part of farmers rather than the optimal decision making assumed by location theory—a sharp rebuke to the latter, therefore—and in part to variations in the diffusion of technical information from major centers like Stockholm and Uppsala.

The work of Golledge (1969, 1970) was also important. Golledge was interested in the process of learning and how the individual's mental map would become denser and, from the learner's standpoint, more accurate over time. One of his early interests was the marketing decisions of hog farmers. Just what was the process through which they decided that one market was superior to another: Greater accessibility? Better prices? Or what?

In conclusion, though, I should emphasize the way in which in behavioral geography most of the underlying arguments and assumptions of the spatial–quantitative revolution remained intact (Cox, 1981; Ley, 1981). Location theory was not being challenged in its fundamentals. Central place theory remained a valid understanding of the distribution of market centers; it was just that some of its assumptions needed more careful specification. There might be deviations from what was predicted, but the underlying logics of market thresholds and range of a good would remain intact. More significantly, there was the idea of some discrepancy between a "subjective" understanding of space—the mental map or the awareness of jobs and housing at different places—and an "objective" understanding calculable in terms of, for example, actual intervening distances, travel times, or transport costs. There are two things here that should be noted. The first is the way in which the idea that the world is out there available to the senses in an unmediated form persisted; this was a central assumption of the spatial–quantitative revolution and relates to positivist foundations, which I explore later. Second, and intriguingly, it is professional geographers who are the discoverers and guardians of this objective knowledge. They are the ones who create the "real" maps. In other words, the idea of expertise lurks in the undergrowth of the mental map literature; something technical, divorced from values, available to "science."

Spatial Determinism

One of the criticisms of the spatial–quantitative revolution was that it was deterministic. In that regard, it might be viewed as a new determinism to replace the environmental one. But instead of the environment determining human activities in different places, the determinism was coming from the pressures of the market transposed onto space: the idea of convergence on minimal cost locations, as in Weberian location theory, or on points of maximum accessibility, as in central place theory, or simply the allocation of different agricultural activities according to the rent that the landlord could extract, a rent that would depend on location. Likewise, migrants moved up interregional wage gradients until equilibrium was achieved.

The charge that such analyses are deterministic is a little unfair. It was recognized early on by spatial–quantitative geographers that there would be deviations from the predictions of location theory and that these deviations might have perfectly reasonable explanations: not in terms of behavioral geography but in terms of other causal conditions having to do with real distributions in the world. Accordingly, one might reasonably expect deviations from the locations of towns predicted by central place theory as a result of deviations from the isotropic plain—a plain across which people could move at equal cost per mile in all directions.

One way around this problem was to treat locations as chance events: events that might have occurred, with a given probability of p, but might not have (with a probability of $1 - p$). This conception led, in turn, to the insights that stochastic models might bring to an understanding of locational outcomes (Curry, 1966). One particularly powerful application of this idea was through what is known as Monte Carlo simulation. The goal here was to simulate a geography as it changed over time on the basis of certain rules about the probability of different events occurring. The simulated geographies would then be compared with what actually happened to check to see if the rules governing the simulation might actually be the ones operating in the real world.[19]

The most noted exponent of this approach was Swedish geographer Törsten Hägerstrand. His work was particularly important in shedding light on chain migration (1957) and on the clustered way in which innovations diffuse over space (1965, 1967). Dick Morrill (1965a, 1965b) adapted these techniques and did further important work on such processes as the peripheral expansion of cities: Why doesn't expansion proceed in all directions at the same rate? Why does new development leapfrog over the existing boundary of the built-up area?[20]

An alternative approach to the problem of determinism has been to recognize the significance of geographic context. Physicists and chemists can control the variables and conditions under which the variables are interacting experimentally and so can establish laws of a deterministic character: Given such and such, then Y will necessarily follow. This is not something that is possible in the social sciences, including geography. So models like Christaller's central place theory or the gravity model have to be fitted to empirical data if they are to be tested. But when that is done, one quickly realizes, for example, that the slopes and intercepts of the relevant regression lines vary from one test to another and that what is affecting them are features of the geographic context in which the models are being tested. Population density makes a difference to how close central places distributing a particular set of retail goods and services will be to each other, as Berry (1967, Chapter 2) made clear quite early on. Likewise, the so-called distance exponent in the gravity model varies according to such contextual features as, again, the distribution of potential destinations (the further apart they are, the lower the distance exponent) or how peripheral a point of origin is with respect to the set of destinations.[21]

CONCLUDING COMMENTS

In retrospect, what is one to make of the spatial–quantitative revolution? As I have tried to emphasize in this chapter, its influence has been huge, if not always in ways that were intended or even desired. It has been the major watershed in anglophone human geography in the 20th century. It drew a line under the dubious legacy of the human geography that had dominated in the first half of the 20th century, a geography that had been backward looking in its substantive concerns and for which theory and method had been virtually foreign words. In consequence, it succeeded in bringing the field out of the closet into a much closer relation with the other social sciences. This worked initially through the RSA and then through invitations to conferences and positions on the editorial boards of nongeography journals, something that would increase over time. Human geography was henceforth recognized as having something valuable to say to the other social sciences. From the low point reached following the field's demise at Harvard, it started to improve its position in the academic pecking order. In more ways than one, human geography became modernized.

Another aspect of this modernization was the radically enhanced specialization that geography underwent during this period. This worked partly on the long-recognized division between human and physical geography and partly on divisions within human geography. Initially, it had seemed that the spatial and the quantitative might provide a means of bringing physical and human geography together. Physical geographers were also busy at the time identifying spatial regularity, as in geomorphology's laws of morphometry. Some of the statistical methodology was shared. The return to geometry that Peter Haggett had heralded in his 1965 book promised convergence, as in the book he wrote with Dick Chorley (1969) on networks.[22] It was, however, a false dawn. As the interest in pattern gave way to a more concerted interest in the processes generating it, so the human geographers turned to models largely of economics provenance while the geomorphologists looked in the direction of engineering and hydrology.

Part of the success of the spatial–quantitative revolution—the reason it had such appeal—was owing to the tenor of the times. In the first place this was highly technocratic. One is reminded of the wonderful title of one of Brian Berry's papers: "Cities as Systems within Systems of Cities" (1964), conveying the sense of something complex but coherent, the fundamental connections of which could be identified and so facilitate intervention.[23] The idea of drawing on spatial–quantitative geography to predict flows and so contribute to planning the rapid transformation of metropolitan areas that was then under way was an alluring one.

Livingstone (1992) has suggested that the emphasis on the quantitative also reflected the politically conservative nature of the times: notably the chilling effect of the investigations of the House Un-American Activities Committee and the McCarthyite hysteria. I am not sure about this. At the very least, its effects would be hard to disentangle from the dominance of what Habermas (1970) called technical over communicative reason, and as Alan Wolfe (1981) emphasized, the public acceptance of this domination had much to do with the relative strength of economic growth during that period.

However, from the early 1970s on, spatial–quantitative analysis was on the defensive, and the early promise that it would come to dominate human geography never materialized. Major points at issue included (1) its spatially determinist character and (2) the positivist character of its epistemologies, which tended to marginalize questions of social values and gave spatial analysis a highly empiricist tone. This latter meant a derogation of unobservables and a naïveté as to the role of conceptualization in research. The mainstream of the field, including the critical mainstream, is now occupied by people who have little time for the quantitative apart from including the occasional table in an article. Many of them have little or no idea what exactly they are averse to. Human geography evolved in a way that polarized the field between quantifiers and nonquantifiers, and the latter then succeeded in dictating the terms of the debate. The quantifiers have become a minority, talking for the most part only to each other.

This seems a great loss.[24] Quantitative analysis, often of a very simple and straightforward sort and easily accommodated to the most math-averse individual, can be hugely important in identifying problems, the deviant case that proves the rule, and, in short, identifying what it is that needs to be explained. Qualitative methods, on the other hand, have to be drawn on if one is to make sense of the correlation coefficients, the residuals from regression, or the simple measures of localization of different industries or ethnic groups. How to bring quantitative and qualitative methods into a more fruitful relationship remains a priority in the field. This is an issue I take up in Chapter 6.

The problem was that the spatial–quantitative work in human geography had allied itself with a particular version of social theory that was ill adapted to understanding the social problems of the late 1960s and early 1970s. Questions of inequality—questions that location theory, with its disinterest in the problem of power, was incapable of dealing with—surged to the fore. The way was being prepared for a further stage in the revolutionization of human geography: its encounter with social theory.

NOTES

1. "Especially from the time of the Second World War onwards, the State has committed massive and ever increasing public funds to regional development programs, highway planning and construction, the building of new towns, urban renewal, public housing programs and many other areas of investment. This has had the effect of bringing urban and regional issues more and more into focus, and of conjuring into being a contingent system of analytical, scientific, and policy discourses as the bases of effective political action. It is scarcely cause for surprise, then, that urban and economic geography (as we know them today), regional science, spatial economics, and urban and regional planning began to make their decisive appearance in universities and research institutes just as late capitalism was beginning to get into full swing" (Scott, 1982a, p. 143). On top of that, there was an interest in a more explanatory human geography indicated by the increasing interest in so-called systematic geography in the 1950s. Part of this was a revival of interest in the spatial, so that human geography was already in a sense being softened up by the contributions of people like Edward Ullmann, Chauncy Harris, and Peter Hall.

2. Peter Gould, one of the more zealous of the missionaries, openly admitted that he hoped his book, *Spatial Organization*, cowritten with Abler and Adams, would become the human geographer's equivalent of Paul Samuelson's introductory text, *Economics*. The point is, though, that Gould's hopes for a standard text, setting out principles of theory and method to which all would willingly subscribe, were widely shared. This was seen as the future.

3. The question of positivism is taken up at length later in the book. Here, let it suffice to identify three of its principal claims: (1) the assumption that observation is pretheoretical; theories are developed on the basis of fact, therefore; (2) the belief that science can only come to conclusions about facts and not values; science should, accordingly, be value free; (3) the same methods as are applied to the physical sciences can be applied to the social sciences as well. As seen later, these are all thoroughly contestable claims, though at the time disbelief was, to put it mildly, suspended.

4. Hartshorne would later (1959) revise his views significantly, recognizing the significance of the systematic fields of geography in providing explanation for areal differentiation. But by then Schaefer had made his point and it was too late.

5. The work of Allen Rodgers (1952) should also be noted.

6. "Although firms exist in which all processes are gathered under one roof, specialization has been historically, and is still, a characteristic feature of the industry's organization. Side by side with the factory, in which as many as possible of the processes of manufacture are gathered, co-exist hundreds of small firms, each carrying out a single stage of manufacture. Associated closely with this feature is the small size of the majority of the firms engaged" (p. 62); and "In these localizations the effect of the co-existence of extreme sub-division of processes and the small unit of manufacture is clear" (p. 72).

7. As, for example, in the work of Doreen Massey (1984) on *spatial* divisions of labor.

8. For an explicit and quite early announcement of this, see Watson (1955).
9. I use scare quotes since obviously one wasn't explaining in a causal sense; rather, one was generating a quantitative relationship, which, however regular it might be, could not necessarily be taken as causal. After all, the "dependent" variable, the variation of which one was trying to "explain" through its relation with an "independent" or "explanatory" variable, could change places with the independent variable with no less variation "explained" in terms of the magnitude of the coefficient of correlation.
10. See, for example, Robinson, Lindberg, and Brinkman (1961).
11. So-called deviations from the distance decay rule were to be the starting point for Hägerstrand's impressive (1957) work on migration.
12. See Yeates (1963) and Cox (1965) for applications.
13. An interesting variant on this appeared in the work of Casetti and Semple (1968). In accordance with the wider interest in planning circles in the 1970s in growth poles, they were interested in how to identify them. Their approach was to correlate the growth rates of settlements on the distance from each of a series of N grid intersections. That provided N coefficients of correlation. The most negative was taken as corresponding to the major growth pole in the area (i.e., this was the grid intersection around which places with rapid population growth tended to cluster). Once identified, the residuals for each settlement were computed and the procedure repeated in order to find a second growth pole, and so on.
14. See also Gould (1975) for an application of what he would call "the Curry effect."
15. Albeit representations without the critical understanding of them that was to come later; rather, and as befits the positivism of the spatial–quantitative work at that time, representations were unmediated reflections of the real.
16. Taaffe's (1974) commitment to what he called "cumulative generalizations" is indicative.
17. Some of the old guard never did come to terms with what had happened. The "Obituary for Stanley H. Beaver" (1985), who was a British geographer of some note during the 1940s and 1950s states that "in his paper of 1982 he came out firmly in favor of a geography which 'is about the whole world, about all aspects of the environment, both natural and man-made, and about people' rather than a geography 'enmeshed in methodological quibbles or in techniques that simply tell us what we know already.'" Evidently, he had come to terms neither with the methodological changes that had occurred nor with the shift to the spatial.
18. For a summary of this work, see Gould and White (1974).
19. A crucial qualification here is the word "might": One could never be sure that the rules were the ones describing what actually happened in the production of empirically observable geographic outcomes.
20. There is still considerable scope for applying this method. How, for example, could one explain the unevenness of gentrification? There are often several different neighborhoods around a CBD that might become gentrified, but typically it will be a very spotty process, occurring more much more extensively in a minority of areas that might seem to have similar things to recommend them for it.
21. See Fotheringham (1981a).

22. The way in which John Rayner, a climatologist, and Reg Golledge, a human geographer, combined to apply spectral analysis to settlement patterns was another example. The origin of this was Rayner's earlier work on the development of spectral analysis for identifying pattern in climatological data.

23. On the ideological significance of systems thinking, see Gregory (1980).

24. Compare Allen Scott (2004) with respect to economic geography: "Economic geographers need to recover the lost skills of quantitative analysis, not out of some atavistic impulse to reinstate the economic geography of the 1960s, but because of the proven value of these skills in the investigation of economic data. The steady erosion of geographers' capabilities in this regard over the last couple of decades is surely a net loss to the discipline."

CHAPTER 3

Social Theory and Human Geography: Material Matters

During the 1960s the spatial–quantitative work was clearly where the vanguard of the field was heading. To take a random sequence of issues of a journal like the *Annals of the Association of American Geographers* from that period is revealing. It is the trajectory that is significant: the increasing numbers of papers that clearly bore the influence of the spatial–quantitative revolution. The rise of the spatial–quantitative had not been entirely smooth. There had always been some resistance. But there was nothing to suggest that its growth, even its assimilation of human geography *tout court*, would be so abbreviated. Nevertheless, it was. The world changed, and as it did, the spatial–quantitative work appeared to be less and less fitting.

There were two major lines of critique. The first was the question of social relevance and the ability of spatial–quantitative geography to respond to such challenges as central city decline, air pollution, environmental deterioration, and processes of spatial exclusion. Initially, this would result in a resort to Marxism as a different sort of approach to understanding in human geography. The second line of critique focused more on the deterministic elements of spatial–quantitative practice, its resort to abstraction, and what was seen as its marginalization of the individual. The response led in the very different direction of what came to be known as humanistic geography.

The critiques proceeded along lines of both theory and method; necessarily so, given the way the ontological assumptions about the world at the heart of different theorizations also entailed methodological change. And to be sure, the decisive events leading to methodological critique were not immediately methodological; there was no independent critical reflection on methods, even while there had been grumbling about method for method's sake and also about the obviousness of some results—the "using a sledgehammer to crack a nut" critique. Rather, the decisive events had more to do with the way geographers' substantive focus started to change in response to what was happening outside the academy. The spatial–quantitative revolution had been appropriate to a technocratic age, but that mood was now being called into question.

Theoretically the apparatus bequeathed by it seemed utterly inappropriate. Making human geography more "socially relevant" in the terms in which "relevance" was coming to be defined meant new theories. Instead of the efficiency question that had been the focus of much of the spatial–quantitative work, human geographers started to worry about inequality: uneven geographic development and the inequality of people's living places and of exposure to environmental hazard. This meant entirely different theoretical priorities. The new questions clearly had to do with politics and power, but the idea of power relations was something that location theory was silent about, something it even denied. Beliefs also started to enter the picture. If people tolerated inequality, and there was certainly plenty of evidence that they did, then what did this say about their views of the world? It seemed unlikely that the economist's idea of a trade-off between preferences for leisure and work would provide the answer.

There were other, quite different questions that, again, the theoretical apparatus of spatial–quantitative geography was ill equipped to deal with. For the most part, and in abstract terms, these had to do with the relation between individual and society. Concretely, this was typified by the planning battles of the 1960s around urban renewal and highway construction. This was construed as a struggle between individual and society, between the concrete person and the abstractions wielded by the state in pushing its agenda. Furthermore, this was seen as a relationship in which the spatial–quantitative geographers were complicit, and on the wrong side. What was called for was an individual-centered human geography that would mobilize on its behalf the various philosophies of meaning.

Both responses to the changing circumstances of the time, that of Marxist geography and humanistic geography, raised issues of a methodological character, issues that would eventually be seen as attacks on the prevailing positivism. Nobody had talked about positivism hitherto, and spatial–quantitative geographers were surprised to learn that that was the methodological altar at which they had been worshipping. But it was to become significant in the debates of the 1970s, particularly on the part of humanistic geography for which positivism became a bête noire.

Marxist geography's critique was more subtle and clearly subordinated to its underlying understanding of the social world and what material practice entailed in coming to terms with that world.

These issues of theory and method form the focus of this and the next three chapters: the core of this book. In this and the chapter to follow, I concentrate on questions of theory and in particular the engagement with social theory, specifically in its more critical forms, that unfolded as a result of the challenges to the spatial–quantitative revolution. In the first place, these had to do with the materiality of social relations—hence the title of the current chapter. In the second place, there is a greater concern for questions of meaning. This would initially take off in the interest in humanistic geography and then take on a much more ambitious form in the embrace of postmodernism and then poststructuralism and postcolonialism, as discussed in the next chapter.

CONTEXT

To understand exactly why the spatial–quantitative revolution came under such vigorous attack toward the end of the 1960s, it needs to be placed in its own social context. Quite why it triumphed in the way that it did is a complex story. There was certainly a sense of crisis within the field itself. The closure of the Department of Geography at Harvard had been a serious blow. According to Neil Smith's (1987) recounting of the events, geography's lack of legibility had been a serious problem for it. Among other complaints, the idea of being the field that synthesized the human and the physical mystified the outsider rather than impressed. From that standpoint, redefining the field around the problematic of space was certainly attractive.

Furthermore, and particularly for human geography, the spatial–quantitative emphasis echoed the technocratic mood of the times. It also provided a new, more applied niche for the field. Henceforth, human geography could respond to the sorts of problems governments were having to deal with in the planning field through its own newly discovered technologies, as Harvey suggested.[1] The way in which the new spatial–quantitative geographers participated in the annual meetings of the RSA and contributed to its publications supported this view, since the RSA had a strong applied emphasis.

There was also a mood at the time that ends were no longer in dispute. Rather, what had to be determined were the means for achieving those ends. In a way alien to people today, it was a period in which mass publics and elected officials were intoxicated by technique, both physical and social. The business cycle seemed to have been mastered through pump priming. The mass unemployment of the 1930s was now history and growth was proceeding at a happy clip. The pressures of the civil rights movement were in the process of being assuaged by new government programs. And the Americans had managed to land the first person on the moon. It was a period that celebrated *The End of Ideology* (Bell, 1960). The social problem of the 1930s seemed to have been solved. Now it was just a question of knowledge. Even as this claim seemed increasingly in doubt as the social question surged once more to the fore, people retained faith in the knowledge that academics could bring to bear on policy.[2] In short, the spatial–quantitative revolution fit the technocratic tenor of the times nicely.

But the dam was about to burst, and by the late 1960s the broader social context and climate of opinion, both academic and lay, was clearly beginning to change. The 1960s had been a period of growing social unrest: the Vietnam War, urban riots, and the challenge of the civil rights movement—the emergence of a clear urban problem in the cities of the United States, particularly in the Midwest, Northeast, and along the West coast. There was also growing awareness of some of the unintended consequences of capitalist development: what would come to be known as "externalities." Rachel Carson's highly influential *Silent Spring* appeared in 1962. There was an emergent critique of economics. Edward Mishan's work was especially noted. In the economics literature, the idea of externalities went back to the work of Pigou in the 1920s but had never been seriously incorporated into economic analysis. This was now to change, particularly in the context of the increasing exploration of the city by economists. A major vehicle for this was the journal *Urban Studies* founded in 1964. Finally, the 1960s was a period of growing confrontation in cities between planners and residents around urban renewal and the rapid expansion of the freeway system. This was especially the case in the United States. Urban renewal particularly affected areas of black residence to the point that it attracted the sobriquet "Negro removal," so subsequent conflicts tended to reinforce a sense of injustice that had been sharpened by the successes of the civil rights movement (Duhl and Steetle 1969). Altogether, the optimism of the 1950s and the earlier 1960s, which had facilitated acceptance of technocratic argument, was fading, and fading rapidly.

The problem from the standpoint of human geography was that the theory that the spatial–quantitative geographers had embraced in their attempts to make sense of the spatial organization that they had pursued with such enthusiasm was singularly ill suited to coming to terms with these social changes. It should be emphasized that the spatial–quantitative revolution did have a social theory. This, in fact, was part of its novelty. Until the spatial–quantitative revolution, the social led a very subterfuge existence. There were some inklings of its importance, as in Mackinder's discussions in *Democratic Ideals and Reality* referred to in Chapter 1 and

Sauer's "culture." But the social had not been something geographers had addressed as a necessary window on human geography. Rather, the focus was on how people related to the natural environment regardless of any social relations that might mediate that relationship.[3] This started to change with the spatial–quantitative revolution. But it was a very limited sort of engagement that could handle only with difficulty the new intellectual challenges. It appeared in two different guises. In terms of influence, there was a dominant one: location theory. There was also a minority interest, which took its cues from the work of Hägerstrand on what were essentially communication networks.

Social Theory and Spatial–Quantitative Geography

The social entered into the theoretical corpus of spatial–quantitative geography in two distinct ways, therefore. The dominant theme was that of the market relations emphasized by location theory. The central focus on location led to an emphasis on the spatial character of markets and hence on competition over space. The competition of suppliers for the orders of firms gave geographically closer suppliers, benefiting from reduced transport costs, an advantage. Migration was now interpreted in terms of a labor market characterized by geographical differences in the scarcity of labor and, therefore, in the wages that workers could command. These differences set in motion a movement of workers from low-wage to high-wage areas until relative scarcities were equalized between the regions and a spatial equilibrium was achieved. In understanding the relation between population densities, land values, and distance from the CBD, attention focused on the structuring of land markets by so-called bid rent curves. Poorer people were willing to pay more for land closer to the (employment concentrating) CBD in order to minimize their transport costs; but given their lower incomes, this meant reducing the amount of land they occupied. Wealthier people who were more indifferent to transport costs were less willing to pay those prices and so located toward the urban periphery, where demand for land was reduced and one could purchase more of it for less. This produced the characteristic decline in population densities and in land values per unit area.

This was the dominant theme in the relation between the social and the spatial–quantitative revolution. There was a more subordinate theme, in fact a very subordinate theme, that emphasized the relation between locational or spatial patterns and the circulation of information. This provided an important condition for the development of behavioral geography. The proposition was a simple one. What emerged in the form of a particular spatial pattern clearly depended on people making decisions: decisions of buyers and renters on where to live, who to buy raw materials or parts from in the case of an industrial firm, and decisions of workers on where to work. These decisions, however, depended on the information that they had about the different alternatives. In its deductive models, neoclassical economics and hence location theory emphasized the perfect information of decision makers, but this was clearly a gross oversimplification. Information had its own geography. Furthermore, it circulated, and in that regard too it might be spatially structured in some way.

The major name here is undoubtedly that of Swedish geographer Törsten Hägerstrand. He spotted the significance of differences in information to the decision process and the way in which it might have some sort of regularity in its geography.[4] His (1957) work on migration departed from a simple observation about migration fields: that they didn't necessarily conform to the distance decay rule of the gravity model and were often characterized by sharp directional biases. Hägerstrand's answer was the variability of the information people had, a variability that did not necessarily follow rules of distance decay: In other words, geographies of communication might be orthogonal to the laws of location.[5]

His work on the diffusion of innovation was also pertinent, and the interpretive framework organized along very similar lines. This led the way for others to develop his ideas. Larry Brown (1968, 1975, 1981) stands out since he took it far beyond the point at which Hägerstrand had left it. The more general notion of imperfect information and location was also notable in the work of Julian Wolpert (1965) on migration, though in a more theoretical context than in the form of empirical work. Finally, we should recall the way in which the idea of some spatial structuring of information found its way into studies of voting geography carried on at that time (Cox, 1969a, 1969b, 1970, 1972b).[6,7]

In sum, the social had finally made its entry into the thinking of human geographers and substantially enlarged understanding of the spatial order at the heart of the concerns of the spatial–quantitative revolution. Even so, in the particular form in which it was being defined, it left a good deal to be desired and its shortcomings were to play a significant role in the further theoretical development of the field. A more complete, thoroughgoing exploration of the significance of the social for human geography had yet to make its appearance.

There were several difficulties—difficulties that, as would later become apparent, were interconnected. The first had to do with the welfare criteria of location theory. These would become of more pressing interest as the 1960s turned into the 1970s and geographers began to express concerns about the social relevance of the dominant location theory, for location theory emphasized a very narrow welfare concern: that of spatial efficiency. The geographic distribution that was efficient, that was produced by the perfect competition assumed by location theory, maximized the benefits to society while minimizing the costs to it. Questions of inequality, variability in welfare outcomes across neighborhoods or regions, issues of air pollution, and all those other things that would eventually fall under the purview of a more socially relevant geography were of no interest. The movement to spatial equilibrium induced by perfect competition would produce an efficient outcome: In response to wage differentials, workers would move where they were most wanted, where labor was relatively scarce, and where the gap between marginal cost and marginal product was therefore greater, in other words; and they would move from regions in which their marginal product was lower and where, therefore, from society's standpoint their labor was needed less. Work that proceeded from the assumption of some spatial structuring of information flows was even less helpful since it had no interest in welfare outcomes at all, whether ones of efficiency or equity.

The second difficulty was the question of power. Power seemed to be significant in the structuring of social relations. Power was a part of lay vocabulary. There were university departments of political science whose raison d'être was exploring the sources of social power. But in the work of the spatial–quantitative geography of the time, it found no echo. Most significantly, location theory could find no room for it. This was because of its roots in neoclassical economics. Neoclassical economics made the assumption of perfect competition. As a result of this, it seemed that exchange was an exploitation-free zone. Sellers trading off one buyer against another and buyers trading off one seller against another were the ways of perfect competition. This precluded the possibility of people exercising power over others in the exchange relation. Everybody, it seemed, was equal. Accordingly, there could be no power differentials of significance to questions of location. In the case of diffusion theory, on the other hand, there was no reason based in theoretical assumptions that would exclude power from logics of innovation adoption. But it was excluded anyway.[8] Attention was focused single-mindedly on the spatial pattern of adoption and not on what the subsequent unevenness might imply about differentials in market power and hence the ability to make the outlays for the innovation or about the very reason for adopting in the first place (i.e., to secure market advantage over others).

Third, the theory mobilized by the spatial–quantitative geographers was ahistorical almost to a fault. This meant that it had serious difficulty in addressing the specificity of the socioeconomic conditions under which the new human geographies of contestation were unfolding at the time. This could be taken as given for anything involving the gravity model; it seemed to apply regardless of historical epoch, as in Tobler's "Cappadocian speculation" (Tobler and Wineburg, 1971). Social relations did not enter into the matter. More complex was the embrace of relations of exchange. The problem here was that markets, the simple business of buying and selling, secreted only a very weak historical dynamic. Markets might deepen, new products might appear, changes in transportation might have effects on the spacing of market centers with respective sets of functions, but quite why that might happen remained outside the theory and reasonably so. The fundamental problem was that the sphere of production remained external to this particular incursion of the social. Its exclusion meant that the impetus to change coming from deep conflicts of interest between employers and wage workers would go unconsidered. Human geographies might move from one position of spatial equilibrium to another, but quite why new equilibria would be necessary was not part of the problematic.

Finally, there was a conceptual separation of the individuals who were doing the locating, and the set of social relations within the context of which they were thus active. For location theory society was determinant of action: locators relocated down interregional price gradients until a spatial equilibrium was achieved, at which point the displacement was brought to a halt; market thresholds and the range of a good would determine retail success or failure. The world opened up by Hägerstrand was, to some degree, at the other extreme. Society was something one engaged with as one saw fit, as in the case of people deciding to move on the basis of information provided by those who had migrated ahead of them or in the case of those deciding to adopt an innovation on hearing about it; or engagement was a chance matter—a letter from a relative at a time when one just happened to be looking for a job. So social relations were indeed recognized as both limiting and facilitating, but just not in the same theoretical understanding.

Critique

By the end of the 1960s the tensions between these assumptions and the world as it was by then being experienced were coming under significant strain. In Harvey's famous words: "There is an ecological problem, an urban problem, an international trade problem, and yet we seem incapable of saying anything of depth or profundity about any of them. When we do say something, it appears trite and rather ludicrous. In short, our paradigm is not coping well" (1973, Chap. 4). The paradigm to which he was referring was, of course, that of the spatial–quantitative revolution.

What was known as "socially relevant" geography was the initial response to this challenge, but it was no more than a way of labeling the transition to an understanding of human geography more alert to power relations and to social theory of a more critical sort. The major innovators were two people who had been close to the core of the spatial–quantitative work: David Harvey and Julian Wolpert.[9] Initially, both developed a critique that was essentially internal to the spatial–quantitative work: a matter of adding new variables like power and culture to location theory. The most important figure here was Julian Wolpert, partly because of the way in which he hewed to this particular approach for some time, while Harvey was to move on very quickly to a critique of a more external character that challenged the most fundamental assumptions of location theory.

Wolpert's (1970) starting point was an empirical observation: Processes of location did not occur smoothly. There were tensions often accompanied by conflict. People resisted some of the locations planned by others, whether private organizations or the state, though he was particularly interested in the latter. In subsequent conflicts the antagonists drew on capacities that were often of unequal magnitude and that would tell—or not, as the case might be—in their favor. In other words, power played a part in location. Location theory wasn't thrown out; it simply needed to recognize the significance of another variable, much as Hägerstrand had argued for deviations from perfect information. Nevertheless, it was an important intervention. It lent some theoretical bite to those interested in making human geography more socially relevant by bringing in the idea of popular resistance. It also stimulated studies of what came to be known as locational conflict. This, along with the work on the geography of voting discussed previously, was to be important in the revival of political geography,[10] and it would become absorbed into what would emerge as welfare geography.

The road that Harvey would very quickly take subsequent to what he called his "liberal" phase would turn out to be the more revolutionary one: an approach, in other words, that radically questioned basic assumptions and saw no possibility of tinkering in order to improve the prevailing theory and make it more adequate to the world. His critique had a distinctly Marxist flavor that would set in motion an equally distinct Marxist geography[11] that has been highly influential in the field.

Despite its enduring influence, Marxist geography would turn out to be a minority affair. It would become something to pay attention to but always in a context of reservation. Theoretically, there was no substitute for it during the 1970s but empirical work would remain stuck in more liberal[12] understandings of the world. Power relations were now central to work in human geography but not always understood in terms of the priority that Marxist geography accorded to the world of production. The state now assumed increased importance, and the emergence of a new political geography was evidence for that. Only toward the end of the decade would something emerge that seemed to provide a more serious theoretical challenge to Marxist geography. This was to be that amalgamation of critical theory and spatial concern that went under the heading "society and space." Just as spatial–quantitative geography was in the process of being forced into a closet, that too would turn out to be the fate of Marxist geography. "Society and space" would be an important step in that direction.

A MARXIST HUMAN GEOGRAPHY

David Harvey, Geographically Uneven Development, and Urbanization

The history of Marxism in anglophone human geography is absolutely inseparable from the name of David Harvey. Through his development of a Marxist human geography or what he later called historical geographical materialism he not only theorized human geography in an entirely different way, but his critique was also a methodological one because, as he pointed out, particular assumptions about the ways of the world entail a particular methodology. Epistemology and ontology were two separate foci of philosophy but Marxism pointed in a different direction: one in which processes of discovery were part of a broader set of social practices given by the social status quo, so that if one wanted to critique the social status quo, then one needed entirely different methods than those that had been drawn on hitherto. Among the methods that had been drawn on hitherto were those of the spatial–quantitative revolution.

In prefacing a discussion of Marxism and geography, it is useful to recall how Marx sets out to introduce his object of study in the first volume of his masterwork *Capital*. His point of departure is the commodity and, therefore, commodity exchange. In a sense, this was nothing new to human geography, just as it was nothing new to those, like Adam Smith, at the center of Marx's critique of political economy. The economic and urban geography that were at the heart of spatial–quantitative work had based themselves on the idea of markets. What was crucial for Marx, however, was the way in which, at a certain point in history and *per necesita*, all the inputs into the labor process, including labor power, became commodities. Accordingly, production and its social logics would change. This was the essential observation. Once immediate producers were separated from the means of production and had to enter into a wage relation with those who had the money to purchase both the means of production and labor power, the curtain rose on capitalism; history and, as Harvey would make clear, geography would be changed forever.

Having laid out money, erstwhile capitalists now had to retrieve it. But there was little point to the exercise unless they were able to impose terms on wage workers that allowed them to extract more value from them than the sum of their wages. They *could* exploit workers in this way since workers could not afford to wait for the equivalent of their means of subsistence. This did not mean that the surplus value extracted would all be consumed by the capitalist class. Some portion of it had to be reinvested along with the values originally laid out and retrieved. Accumulation was a necessity since capitals found themselves in a relation of competition with other capitals. Only by expanding and so increasing the flow of surplus value, which could then be used to drive out the competition through a variable mix of takeovers, exploitation of economies of scale, the development of new and improved products, and increased capital intensity of production, could survival be assured. And all the time a new world was being created on the backs of the immediate producers.

Initially, the application of Marx's thinking to human geography was more in line with the substantive emphases of the spatial–quantitative geographers: urbanization and suburbanization and regional development. But now it was from a standpoint that was sharply critical of the priority that had been given to commodity exchange. Geographers began to penetrate what Marx had called "the hidden abode of production," demonstrating the sharp inequalities that characterized it and what those inequalities portended for the creation of other sorts of inequality, including those between people in different places. At the same time, it introduced a welcome dynamism into thinking about the changing space economy. Not only was the map of uneven development a shifting one, responding to the dynamics of the accumulation process, but there were tendencies rooted in the capitalist mode of production toward the suspension of the barriers of distance. Capitalism had it within its capacity to revolutionize not just social relations but space relations as well.

We can get some sense of what all this implied for the study of human geography from one of Harvey's earlier articles from that period. "The Geography of Capitalist Accumulation" appeared in 1975, and what is remarkable is how much that is fundamental to a Marxist geography was brought together by Harvey in that early effort. Class relations remain implicit in the analysis. The focus is on accumulation and its geography, yet given the class relation as the condition for accumulation it can't be far away. He starts out by identifying the central role in any social process, including spatial ones, of contradiction. Capitalist development proceeds by creating barriers to its further development, barriers which it then suspends but which will later reappear, perhaps in another form. It is contradictory in the sense that its logic constantly threatens its own reproduction. Accumulation means producing a surplus, but then that surplus has to be disposed of somehow. Capitalists are not going to consume it all, since if they are to remain capitalists they must grow; part of the surplus must be reinvested. The question is, in what? One possibility is excluded by the logic of capital: The surplus cannot be shared with the workers since then capital's ability to make a profit would be threatened. There are others, though. Capitalists can create new wants through the development of new products. They can also open up new spheres of activity to capitalist production, as in replacing peasant production with that of farms devoted entirely to production for the market. But both of these may entail expansion of a geographic sort: procuring raw materials for new products or extending capitalist production into other parts of the world where precapitalist forms of production continue to dominate.

This impetus to geographic expansion entails revolutionizing transportation: new modes, improved modes. Transportation is an added cost of production. The firm that can lower its costs of transportation will be a more effective competitor. The drive toward successive revolutions in transportation—the age of canals, the railroad era, steam navigation, and still later the container revolution—is on. Inherent in capital is the abolition of spatial barriers. These are not just barriers of cost, for time is also significant. Commodities in transit tie up capital: The quicker goods reach their destination, the quicker they can be sold and the money reinvested in the expansion of production. The agglomeration of producers has similar implications. In short, the circulation of capital is subordinate to production.

The process is driven by contradictions but throws up new ones. To overcome the barriers posed by space, capital is invested in new spatial structures: railroad lines, dock facilities, the physical infrastructures of cities. But these then become a barrier to the further attenuation of spatial barriers. Railroads, canals, tunnels, electricity grids, port facilities, and urban infrastructures represent massive investments of long life; their cost has to be amortized over many years, possibly two decades or more. This means that the geography of movement must continue to conform to the existing geography of fixed investments for some time into the future. In other words, the further tearing down of spatial barriers has to be put on hold. To paraphrase Harvey, capitalist development has to negotiate a knife edge between preserving the values of past capital investments in the built environment and destroying those investments in order to open up new room for capital accumulation. In short, the space economy is far from being the expression of harmonious equilibrium portrayed by the location theory of the time; rather, it is an arena of contradiction and tension.

This hints at the possibility that the world of capitalist production has important ramifications for politics and for political organization: negotiating that knife edge that he refers to through resort to the state. His discussion of imperialism then makes this connection explicit. The expansion of capital around the world was accompanied by the extension of European rule in the form of empire. Harvey does not go into the connections in detail, but one can see what they might be: notably the need to impose a particular set of property institutions if capitalism is to take root. What is more interesting, though, is how he uses this discussion to underline the open-ended character of capitalist development.

Capitalism has a logic of reproduction on an expanded scale that is hard to turn aside. It subordinates all moments of society—the state, culture, discourse, relations to nature, institutions—to its logic. The more concrete ways in which it does this, though, vary. Nothing is set in stone except that *in toto* the concrete form of a capitalist society has to be consistent with accumulation. Social development is open, but with that crucial proviso. And so it is with imperialism. There have been many, highly specific, Marxist explanations for imperialism. For some it has been a matter of expanding markets for finished products; for others it has been one of finding new destinations for capital export, places in which to build railroads and so find an outlet for a surplus that lacks other outlets; for still others, the crucial consideration has been laying hands on new or cheaper raw materials. But what Harvey emphasizes is that all of these are possible. There can be no single theory of imperialism except to ground it in the contradictions of the accumulation process.

The spatial–quantitative revolution had generated in the mind a particular sort of human geography. For sure, it was one that emphasized the economic, the urban, and transportation but in a certain way: a way whose features were put in sharp perspective by the arrival of a challenger in the form of a geography that took its points of bearing from Marxist social theory. Spatial–quantitative geography abstracted from a wider context of social relations to concentrate on the economic; in treating the economic it found no place for power relations, in contrast to those unequal social relations, which in the Marxist view entailed exploitation; and it had little or no sense of dynamics: It was in thrall to the idea of equilibrium, whereas Marxism, while accepting movements toward that position, saw it as something that would be constantly upset as people developed, though always unevenly (see Table 3.1). Accordingly, the substantive foci shifted. Instead of central place theory and its aesthetic of symmetry and the beauty of spatial organization, attention now turned to geographically uneven development and the tensions through which it got transformed, as in Harvey's understanding of the geography of accumulation. The city was still of interest but the angle of vision had changed quite dramatically. In part, it was the process of urbanization that was emphasized and how it was a vehicle for the absorption of a surplus that was continually threatened with having nowhere to go. In part, the city was now viewed as a contested terrain as capital sought to remake it in ways through which it could be maintained as an infrastructure that would facilitate production and the appropriation of rent; while those threatened in their living places by the new highway construction and urban renewal so entailed resisted in ways that recalled Julian Wolpert's early work but understood within a theoretical framework that was quite dramatically different.

Michael Watts and Political Ecology

If the substantive preoccupation of spatial–quantitative geography had been the city, transport, and the general problem of location, we also

TABLE 3.1. Alternative Approaches to the Geography of Development

Quantitative geography	Marxist geography
Fundamental concepts	

• Taken for granted: combine commonly accepted "indicators" of development into an overall index via factor analysis	• Situated within the sphere of human practice:
	  • Concepts as having a history: a changing set of sense relations
	  • For whom are they concepts? For whom do they work? Why and how?

Time	
• Cross-sectional: comparing one distribution with others at one point in time	• Historical: What are the necessary conditions for "development" and what is their history in a particular instance? How did they get instantiated?
• The system of relations tends toward a spatial equilibrium: capital moves to where it is relatively scarce; workers move to where they are relatively scarce. Once the relative scarcities are eliminated, there are no incentives to further movement because people cannot better the returns to their resources. Geographically uneven development has been eliminated.	• There are systemic barriers to the achievement of equilibrium so that geographically uneven development can never be abolished.
Techniques	

• Quantitative, aimed at the identification of empirical regularities; e.g., "center–periphery" relations ("distance from Nairobi" and development in Kenya):
  • Reduction to a statistical soup in which particularity is lost
	• Historical reconstruction of particular cases:
	  • The interpretation of meaning
	  • In terms of changing class relations and the concrete trajectory of the accumulation process
	  • The identification and exploration of particular strategies

Pluralizing/totalizing	

• Pluralistic:
  • Measuring the effects of different "independent" variables
  • Development as "economic"
	• Totalizing:
	  • "Independent" variables rejected from the start
	  • Development as a matter of forms of production conditioned necessarily by the political, the cultural, etc., and reworked in accordance with the demands of production. Ultimately development is about the development of people as people, allowing the flowering of their human potential.

Universal/particular	

• Universalizing: the search for empirical regularities of increasingly universal scope as in "the results need to be confirmed by further research elsewhere."	• Both universalizing and particularizing:
	  • The particular as an expression of the universal, but not reducible to it
	  • The particular as becoming universal through its internalization by the global logic of capital

know what it was not. As we saw in Chapter 1, for the first half of the 20th century and perhaps even as late as the late 1950s, human geography had been about the relation between people and what was termed at the time their "natural environment." This particular focus found barely an echo in the spatial–quantitative work of the 196s. In terms of Taaffe's three traditions of geography, attention swung clearly and almost brutally away from that relationship to a single-minded focus on the spatial, or, more accurately, the sociospatial in abstraction from the ecological. Only toward the end of that decade did the people–nature relation begin to put in an appearance again. This would be in the form of hazards perception research and how perceptions mediated between the hazard and human adaptation.[13] But like the spatial–quantitative work, it lacked any sense of power relations, and therefore, of the social relations grounding them. Its impact on the field was quite limited. This would definitely not be the case with what would come to be known as political ecology. Now people–nature relations would be placed on a much firmer footing. The fundamental relation was still there, but now it was mediated by social relations and, in its initial incarnation at least, by the social relations of production at the center of Marxist theorizing.

Harvey had set out the problematic of what would become political ecology in a 1974 paper attacking neo-Malthusian arguments about the population–resources relation that were currently *au fait*. It was, however, Michael Watts who would become most strongly associated with political ecology. His book *Silent Violence* (1983c) on famine in the Sahel and an article related to the same body of research (1983a) lays out much that was apparent in Harvey's earlier contribution and that has become the received wisdom for many in political ecology. His counterpoint, however, is more clearly the natural hazards research of the late 1960s and early 1970s. He makes the fundamental point that the vulnerability of rural producers in northern Nigeria to environmental hazard was a function of the integration of precapitalist forms of production into a global capitalist system largely under the aegis of the colonial state. In other words, the idea of hazards as "natural" had to be subjected to critical scrutiny. What might be a "hazard" under one set of relations of production might not be under another.

Before the arrival of the British colonial state in what had been Hausaland and what is now northern Nigeria, the basic unit of production was the extended household. Each household belonged to a fief and the lord holding the fief appropriated surplus labor in the form of labor services, grain, or cash. Drought had occurred in the past and there had been famines. But peasants had been able to adapt by, for example, spreading their risks by planting a diversity of crops with different moisture needs and redistributing food through kinship. In addition, at different levels in the political hierarchy, grain was stockpiled as a guarantee in case of famine. As Watts observes from his research: "Peasant security, and by extension the hazardousness of place, was intricately bound up with the nexus of horizontal and vertical ties which were coterminous with the social relations of production characteristic of a tributary system of production" (1983a, p. 30).

The colonial state altered these arrangements quite drastically and created conditions that would later be identified, wrongly, as risks of a "natural" sort. Colonial taxation meant that peasants now had an incentive to produce for the market; what would seemingly fetch the best price came to dominate rather than the old strategy of planting crops with different moisture needs. Transfers to traditional rulers came to an end and with them the stockpiling of grain as an insurance against food shortage. The Hausa state had been flexible in its demands of the peasants. In times of shortage, its appropriations of the surplus would be reduced. In contrast, colonial taxation was rigid and took no account of poor harvests. Likewise, taxes were often due immediately prior to harvest so crops had to be sold when prices were low or, alternatively, fall into the hands of money lenders.

As Watts states, and contrasting his approach to that of the natural hazards school:

> Natural hazards are not really natural for though a drought may be a catalyst or trigger mechanism in the sequence of events which leads to famine conditions, the crisis itself is more a reflection of the ability of the socio-economic system to cope with the unusual harshness of ecological conditions and their effects. To neglect this fact is to resort to a fatalism which sees disasters as metaphysical, as acts of God in which the responsibility is shouldered by Nature. In the process, of course, this misses a major political point. The fact that they are not is expressed by the unevenness with which people in the area are affected by the famine: urban people much less so than the rural, for example. (1983a, p. 26)

In developing these points, there are some subtleties. In particular, as Watts argues, the "wider socioeconomic context" has to be specified very carefully so that it is not reduced to the relations of exchange in which producers find themselves; rather, it is also a matter of the social relations through which people relate directly to nature in the production process. Hence, in the case study examined, it is not just the incorporation of Hausaland into an exchange economy that is significant but also the failure to change the production relations so as to enhance productivity. Perhaps the point can be grasped more intuitively by saying that to argue that what has happened is simply an addition of "economic risk" to "environmental risk" is to fall short of an adequate analysis. The economic cannot be separated from the environmental. The relation to nature is always and necessarily a social one, as in relations of property. Economic relations are always environmental ones and vice versa.

In short, the people–nature tradition that Taaffe (1974) had referred to in somewhat dismissive terms in his AAG Presidential Address of 1973 was back. It was back, however, in a very different form from that which he had criticized. The fundamental point was that the people–nature relation was now seen as always mediated by social relations. More specifically, it was the social relations of production foregrounded by Marx that were to be at the center of this new instantiation, though that would change somewhat with the elapse of time.[14] To the extent that social relations were capitalist, then the accumulation process would be the starting point, as indeed it was in Harvey's (1974) treatment of Malthus. More often, however, political ecologists like Watts focused on developing societies where there was some sort of combination of the capitalist and the precapitalist. But to the extent that those relations could be correctly defined and their logics of class relations and reproduction understood, then light could be shed on how people mobilized natural forces and naturally occurring substances to the end of their own reproduction.

Rethinking Landscape

If the significance of people–nature relations had been marginalized by the spatial–quantitative revolution, so too was that the case for the idea of landscape. Recall just how important landscape was to the work of Carl Sauer and his students. Part of the marginalization surely had to do with the way in which Sauer had seen it as a relation between people and the land, but it was also a result of its seeming subjectivity and resistance to quantification. Landscape was now to undergo something of a revival, albeit as part of a revivified cultural geography that was only partly conditional on the interest in Marxism. It is, however, the Marxist work on landscape, associated in particular with people like Denis Cosgrove, Stephen Daniels, and Don Mitchell, that I want to emphasize here.

For Sauer, landscape had been something that was naïvely given; it spoke for itself and was the expression of a particular culture. Landscapes expressed values. They could be "good" or "bad." A "good" landscape was one that exhibited a harmonious relation between people and nature: one in which local materials had been used in the construction of houses and the divides between the fields, in which roads conformed to the lie of the land. In "bad" landscapes this sort of organic relation was lost.[15,16]

One of the crucial innovations of the Marxist revival of landscape studies was to recognize that far from something that was naïvely given, landscape was an historically given "way of seeing": a way of perceiving structured by social struggles that were concealed by this particular way of representing the world and in utter contrast to the homogeneous and seamless "culture" assumed by Sauer. According to Cosgrove (2006), the "landscape idea" was "a characteristically modern way of encountering and representing the external world: in its pictorial and graphic qualities, in its spatiality and ways of connecting the individual to the community. As reflected in diverse forms—works of art, the images in coffee table books, picture postcards, architectural design, and land use planning—as well as in everyday experience, landscape answered to human needs born of the alienations of capitalist society, and it provided an ideological fix for the challenge to the status quo that those alienations might generate.

On the one hand, therefore, landscape banished the world of production and provided a compensation through images of something that was the very opposite of that world, as in the aesthetic of the suburb, the country seat in a park-like setting, or the country retreat visited on the weekend and on holidays. This was indeed a compensation for the alienating experiences of the workplace: images of spaces organized exactly for leisurely consumption and not for production, spaces in which all traces of functionality, all traces of the world of production, were excluded.

On the other hand, it was an ideological fix. This was because of the illusion of harmony and balance between people and environment that it provided. It was a framing that excluded the conditions that allowed that harmony and balance to be achieved; the social struggles and exploitation that went on seemingly offstage but without which the particular view of prosperity, orderliness, and artfulness would not be possible. As the title of Don Mitchell's (1996) book conveys, it was a matter of "the lie of the land." Landscape was something that was literally fought over. It was an imposition of power. It depended on the creation of landscapes elsewhere that rated much lower on what Cosgrove (2006) called "a moral barometer": the landscapes of mill towns, colliery villages, the agricultural landscapes of parts of Illinois—no doubts there of being organized for production. As Mitchell (2000, Chap. 5) has emphasized landscape is an imposition of power. The preservation of the suburban aesthetic of tree-lined, curved streets, seemingly custom-designed houses and spacious lots on cul-de-sacs from which all facts of production—stores, gas stations, parking lots, as well as factories, even commercial transport—are excluded, along with any residential uses that would damage the particular balance that is the suburb, like high rises and standardized terrace forms, requires the mobilization of power as in the case of numerous land use controversies. Landscape as "moral barometer" reinforces senses of one's place, quite literally, in the world. Just as employment has to be defined as a superior condition to being unemployed, so the landscapes of the affluent have to be defined as similarly superior: Reproducing a mode of production requires that aesthetic tastes be in conformity with that project.

There is, therefore, something utterly conservative about the idea of landscape. It emphasizes what the capitalist world is clearly *not* about except as a means to legitimating accumulation for accumulation's sake: leisure and aesthetic experience in the case of the landscapes of bourgeois suburbia; landscapes of timeless sustainability where it is a matter of the countryside, existing in stable organic relationship with the people who live there. It can, however, be defined as something that exists apart from the people who live there: something enduring, reflecting the values of a different age for which in a context of constant change, there are feelings of nostalgia. People live in the seemingly quaint and picturesque villages of the English countryside even while the landscape of thatch and local stone is mere shell for interiors of designer kitchens, walk-in closets, and bathrooms with "his-and-her" vanities. As such, it can be mobilized in fights over what from one angle is the country's patrimony—images of William Blake's "green and pleasant land"[17]—and from another, something that enhances immeasurably the very different patrimonies of those who occupy the houses.

Neo-Smithian Marxist Geography

Marxist geography placed production at its center. There was, though, a deviation that concentrated more on commodity exchange. This obtained its inspiration from the work of a number of maverick Marxists and would come to be termed, in a thoroughly pejorative way, "neo-Smithian Marxism." The term originated with the Marxist historian Robert Brenner, who wrote an article in *New Left Review* in 1977 excoriating the tendencies among some[18] to reduce capitalism to market relations. The extension of market exchange on a world basis, it had been argued, led to a geographic division of labor, which worked to the advantage, through processes of unequal exchange, of the more developed countries. The latter then used their political power to enforce that particular geographic division of labor. Thus was created a geography of center or metropole on the one hand and periphery on the other: a conception that readily appealed to human geographers wishing to move beyond the descriptive concepts of what would be known as welfare geography (see below) so as to draw on a process that could be seen to be generating inequalities between places and that also emphasized the relations between places as fundamental to that process of production. Yet, as Brenner pointed out, this emphasis on the exploitation of one place by another ignored the major source of exploitation, which was in the workplace: in other words, the exploitation of one class by another.[19]

Work along these lines continues to endure, particularly in political geography. The most popular of all the introductory textbooks in political geography, written by Peter Taylor and Colin Flint (2007), is based on an application of the claims made by one of the most prominent of Brenner's neo-Smithians: Immanuel Wallerstein, who formulated something called world systems analysis. Taylor, in fact, is the most visible of all the human geographers who have continued to hew to this particular line.[20] In particular, he has been to the forefront in world city research. But now, instead of centers dominating other places, including peripheries, it is world cities.

THE CRITICAL MAINSTREAM

A notable feature of historical materialism and the Marxist geography to which it would give birth is its conception of production. What are typically seen as separate from production, including exchange, circulation, and distribution, are now identified as necessary aspects of production itself. Production in the narrow sense of the labor process predominates and structures the other aspects of production taken *in sensu lato*.

This unity was to be ignored in what I have called the critical mainstream. For appealing as the vision of Marxist geography was to some, for most of those who sought a way out of the seeming social irrelevance of the spatial–quantitative revolution its influence would be more diluted. Whether this was due to the daunting challenge of making sense of Marx in any sense other than superficial[21] or inhibitions born of the discouraging discourse that was concomitant of the Cold War is difficult to say.

This was particularly the case with what I call "welfare geography." This was focused largely on distributional questions: on questions of who gets what, where. It was certainly left leaning and illuminating but lacked explanatory depth. It would eventually be absorbed into what would come to be called critical human geography: a rather amorphous hybrid in which the spatiality provided by the circulation of values would be conjoined with that defined by a circulation of meanings.

More durable and retaining its own identity would be the more critical focus on questions of production that emerged in the wake of criticism of spatial–quantitative geography. This would form the basis for a reworking of economic geography giving it a much stronger spatial focus than had been true of its spatial–quantitative incarnation along with theoretical credentials, even while they did not add up to a coherent theory in themselves, that were more compelling than anything that welfare geography could offer.

Welfare Geography

Welfare geography was emphatically an approach rather than a theory, initially expressed in the early issues of what was then the new journal *Antipode*, which first appeared in August 1969. It was an approach, moreover, that had been laid out by Harvey (1971) immediately before his Marxist turn in a paper devoted to explicating the relation between urban systems and the distribution of real income.[22] It is the closest that one would come to as a manifesto for welfare geography. Certain features stood out, features that clearly demarcated it from Marxist geography.

The first was a concern with the distribution of what, for want of a better term, one might call "well-being." There was an implicit criterion of social justice in the critical mainstream that found fault with existing geographies because of the way in which they tended to work to the disadvantage of the poor: Their share of the pie was disproportionately lower and this showed up in diverse ways—not just incomes but also what could be purchased with them, including housing and neighborhood amenities or the lack thereof. The material oppression of the masses under capital is a major theme in Marx but his emphasis lay elsewhere. It was, rather, the sphere of production and the social relations that allowed production to occur, indeed necessitated its particular social form that were of primary significance. People entered into production as wage workers because they had no alternative; they had been dispossessed of the means of production. But in order that profits be made, downward pressure on wages was a necessity. To the extent that those distributional arrangements were upset, then investment would decline and production would be impaired. This was a vision that was less obvious in its geographic implications than a focus on distributional outcomes, a focus that would later assume the evocative term "welfare geography" (Coates, Johnston, and Knox 1977; Colenutt, 1970; Knox, 1975; Smith, 1977, 1979, 1988) in which inequality was mapped, seemingly *ad nauseam*, and typically in relation to the distribution of income.[23]

Second, markets and their dysfunctional consequences were highlighted. The activities of utility maximizing individuals[24] working under the constraints of market forces could have unintended consequences or externalities for others, some of which might be quite detrimental. There might also be monopoly effects. Both of these consequences could again be given a geographic interpretation. Externalities like air pollution had a geography; one could talk about *spatial* externalities. The same applied to monopoly effects; something that had been picked up on in central place theory but that now received a new, distributional interpretation. Again, markets were a central feature of Marx's understanding of capital. Chapter 1 of the first volume of *Capital* opens with a discussion of the commodity and commodities entailed in market exchange; the idea of the commodity made no sense otherwise. For Marx, though, markets did not have the foundational status for capitalist societies that they would have for geography's critical mainstream in the 1970s. Rather, as I pointed out previously, it was the accumulation process that was central and this was a class relation. Capitalists accumulated through the exploitation of the labor power of the masses. Commodity exchange simply made necessary a relation already implicit in the confrontation between those with money wealth and the dispossessed. In other words, externalities and monopoly were beside the point; the most serious oppressions of the working class were occurring at the point of production.

Third, the critical mainstream seemed to live in a world in which one commodity owner was as good as another, with the critical difference that some had more money or were advantaged in some other way, for example, by location, race, or gender. Rather than class, the central concept seems to have been one of social stratification. Racial or gender difference were imprinted on the landscape. The power of money changed geographies regardless of whether one was talking about those who put their money to productive purpose, as with business firms, or those who put it to consumptive purposes, as in homeownership. The power of money, so emphasized in Marx, now assumed a continuum, regardless of the qualitative difference between money that is thrown into production and that which is devoted to consumption. Again, there was a sense of Marx but without the essence, and this was clearly connected to the distributional emphasis.

Fourth, there was the state. The state was important in Marxist understandings of capitalist society, but this was a different conception. Obviously, the state could not be ignored in any defensible conception of a socially concerned human geography. This, however, was a state that was, in effect, up for grabs: a source of power that was neutral with respect to all the fissures in society but that could be mobilized for narrow purposes. Money wealth had an advantage in tilting the state in its direction. The moneyed could fund the election campaigns of candidates for office. They could hire the legal experts if they didn't have the knowledge themselves, though often they did. They could threaten to take their (taxable) wealth elsewhere, whether it was wealth invested in factories or in expensive homes. Echoing what was said in the previous paragraph, wealth was wealth. And yet again, there are echoes of the Marxist view of the state, but they are very faint ones. For Marx, the state was never neutral. It certainly did work to the advantage of money wealth, but it was the wealth of capitalists, of productive wealth, that held it hostage in ways of which it was aware as capitalists threatened an investment strike as well as unaware, as in the ideological categories into which everyone in a capitalist society, regardless, was socialized.

These major assumptions then fed into a distinctive geographical focus. This was one in which space became a stake in distributional struggle: in which people struggled over the incidence of externality effects, as Wolpert had earlier specified, attempting to attract government support; or in which they were entrapped for reasons beyond their control. The central mechanisms of this geography then became the related ones of locational conflict[25]; residential exclusion; residential displacement; and the way differential mobility could be turned to advantage. All were processes in which wealth counted.

In locational conflict, the more well-to-do were enabled by their knowledge of the political process and the fact that members of the political class were part of their social circle. They were also adept at taking advantage of the law, particularly zoning law, in order to erect barriers to keep the poor out of their neighborhoods: opposing the rezonings that would allow in housing that the less affluent could afford. To the extent that the geography of residential desirability shifted, then money would speak again, bidding up prices in the newly fashionable gentrifying areas or the villages with their attractive rural ambience surrounding British towns and forcing out the poor. The wealthy were more mobile; there were fewer limits to what they could afford in terms of where to live, and they had more cars. As they moved, services like physicians and the supermarkets followed them, leaving poorer parts of the city deprived and, in virtue of fewer cars, in effect trapped in space and subject to the monopolistic exploitation of those retailers who remained. Limits to the expansion of cities imposed by the wealth on the periphery forced up rents for the poor, and so on.

The idea of spatial mismatch was a central theme. Suburban employers needed workers at the same time as inner-city areas were afflicted by high rates of unemployment. In part, this was believed to be due to the exclusionary zoning favored by the more affluent; poor people could not buy or rent housing closer to where the jobs were. In part, it was due to a decline in the availability of mass transit; something entailed by the way in which those with money had favored the auto over the bus. Something known as the metropolitan fiscal disparities problem also attracted attention: In virtue of their relative poverty, central city populations had greater need for public services, but the taxable resources out of which they could be funded had migrated to the suburbs.

The pluralism of this approach was striking, and that was one of its attractions. Focus on markets rather than accumulation allowed a separation of supply and demand forces that could open up explanatory possibilities in new ways, however indefensible they might appear when viewed from a Marxist standpoint. Even humanistic geographers could be part of the project as residents sought to protect particular values of a seemingly nonmaterial sort, and so demonstrating the inadequacies of a supposed Marxist economism.[26]

Critical Economic Geography

A different strand of critique of spatial–quantitative work, including its obliviousness to questions of power, was, like welfare geography, an internal critique, but now focused on logics of production and with a much more enhanced spatial sensibility when compared either with that which it sought to replace or with welfare geography. This was a more serious intellectual contribution with much stronger theoretical roots and sense of cumulative inquiry.

Not least there was a stronger sense of the whole–part relation. This entailed much more attention to questions of the wider political economy and its structuring role. This has meant, among other things, a periodic rejuvenation as the concrete form of that political economy, or at least how it has been interpreted, has undergone change: so from restructuring to regulation theory, and then from regulation theory to globalization and the global–local, though clearly with important overlaps and continuities.[27] It remains an important influence in economic geography, and its practitioners were prominent in the creation of the new periodical *Journal of Economic Geography*. A distinctively new economic geography, quite different from that associated with the spatial–quantitative revolution was, therefore, to emerge.

This was a critique that put at its center the impossibility of the assumptions of neoclassical economics and in particular those that abstracted from the fact of spatial variation. In this it shared something with welfare geography, particularly studies of locational conflict, but with a much stronger awareness of the economics that was being critiqued as well as a stronger record of innovative thinking. This would have important implications for an understanding of geographically uneven development, even while, for the most part, its guiding framework of ideas lacked intellectual coherence and put class tensions in the background. Prominent contributors to this line of thinking have included people like Susan Christopherson, Gordon Clark, Meric Gertler, Ron Martin, Michael Storper, and, most notably and durably of all, Allen Scott. The early Massey shared something with this tendency though with a stronger focus on the question of class.

The precepts of location theory, based in neoclassical economics and vigorously applied in the urban economics and regional science of the 1960s as well as in spatial–quantitative geography were the critical target. This would be part of the ongoing spatialization of social theory, a spatialization that would severely test the conclusions of what had gone before. Most notably, ideas of externalities, transaction costs, immobility, and the agency of the firm would undermine the received wisdom of spatial equilibrium. In turn, this would open the door to claims about how, therefore, the space economy might be regulated so as to counter its clear tendencies to disequilibrium and allow some version of balanced growth.

Scott's entrée into more critical understandings of the space economy took the form of a focus on urban land markets, and in particular the difference that externalities and immobilities made to the way they functioned (Roweis and Scott, 1978). The externalities subsequent to new land use decisions and the slow convertibility of adjacent land uses condemned urban land markets to an inherent instability. A variety of territorial conflicts emerged redolent of the cases that Wolpert had studied but now placed in a broader context of "the urban land nexus."

Later, in the 1980s, and picking up on themes introduced in his book *The Urban Land Nexus and the State* (1980), Scott focused more on the city as a production mechanism: how the socialization of production through a division of labor between firms was simultaneously a spatialization that was a necessary condition for the urban. He was particularly interested in the vertical disintegration of firms as a condition for their agglomeration (Scott, 1982b). He picked up on the work of Coase and Williamson on transaction costs and the implications of the latter for firm governance. The argument was that if transactions were externalized, then the costs of securing information about clients and their products, their reliability and trustworthiness would be proximity sensitive (Scott, 1985). Clark and O'Connor (1997) would also draw on the idea of transaction costs and their implications not for urbanization but for the hierarchical organization of the space economy. Different sorts of financial product were more transparent than others. Where transaction costs were significant in their purchase—that is, the informational content of financial products was substantial—then one could anticipate more local markets as financial service firms sought to take advantage of local knowledge.

Clark was also to the fore in the early restructuring literature. A notable contribution was his (1981) work on the employment relation and its implications for firm spatial divisions of labor: a nice forerunner of Massey's seminal work and in some ways more satisfying from the standpoint of mechanisms. His emphasis was on segmented labor markets and the reasons having to do with wage and benefit bargaining for geographically separating the less skilled parts of the labor process from the more skilled. His work on sunk costs (1994; Clark and Wrigley, 1995) and the problem of immobility for firm restructuring should also be noted.

The theme of restructuring[28] focused on the firm though in the form of (the British) locality studies (Cooke, 1989; Harloe, Pickvance, and Urry, 1990), local governments were also brought into its embrace. But given the interest in the impossibility of equilibrium in the space economy, complementary themes of regulation and, therefore, the state could never be far away. This was apparent in Scott's early work on urban planning and what he called the urban land nexus (1980). It would be later taken up with vigor as economic geographers discovered regulation theory. Once again, some of the more innovative work came from Scott and his attempt to give locational expression to what became known as flexible accumulation. This would be his work on new industrial districts, building on earlier claims about the role of transaction costs in agglomeration processes (Scott, 1988). Given the sparsity of examples of new industrial districts (Lovering, 1990), this pushed the envelope,[29] but it did capture something of what was happening, as did Storper's related work (1997a, 1997b) on the untraded interdependencies of firms.

Possibly the most important and enduring effect of the interest in regulation theory was what became known as the institutional turn in economic geography. This would consider the role of not just state institutions but also those that had emerged in the form of country-specific ways of doing things: of corporate organization, of the employment relation. Some of this came about in critique of regulation theory; how, in the case of Sayer's important intervention (1989b), Japan was different. This would then shed a new and critical light on some of the more optimistic views of globalization. Both Gertler (1997) and Christopherson (2007) would produce important case studies on the obstacles to transference of technologies, physical and social, out of the particular institutional contexts that had given birth to them.

These lines of investigation taken together have formed the core of a new economic geography. Unlike the old one, this is an economic geography much more attuned to the problematic of uneven development, as is discussed in Chapter 5. On the other hand, the spatialization of social theory continues, most recently in the form of work on path-dependent development (Martin, 2010a, 2010b). The conclusion is that development is not just path dependent, but it is also place dependent. The conditions for this conclusion go back to Scott's earlier work on transaction costs and agglomeration and what agglomeration suggests about increasing returns to scale and monopolistic competition, something that he had noted early on (1982b).

In hindsight, and confining ourselves as per the title of this chapter to "material matters," the new critical economic geography has been the most enduring and influential of all the original critiques of the spatial–quantitative work. Welfare geography as I described it has also gained a wide constituency but without the same intellectual heft. From one angle this success is odd. It is a critique of that economics mainstream that informed the economic geography of the 1970s and a drawing out of its spatial implications, but that is about it. Unlike Marxism, it has no clear theoretical coherence: no alternative vision of the world other than that in some details orthodox economics and, therefore, location theory got it wrong. The results have certainly been insightful and added immensely to our ability to make sense of the shifting space economy, but the failure to cast its claims within an overall critique that puts all the assumptions of that orthodoxy in doubt continues to nag.

It has been the great survivor, though. In contrast, Marxist geography's star has waned. Its promise as the backbone of a new critical human geography remains unfulfilled. Even the journal that was its mouthpiece turned significantly against it, resulting in the creation of a new one, *Human Geography*. What initiated that gradual process of closeting, at least at the level of ideas, has to start with the couplet "society and space," which began to appear with increasing insistence from the end of the 1970s. An important point of continuity, though, would be feminist geography, to which we turn first.

FEMINIST GEOGRAPHY

All the critical developments mentioned so far would be reflected in a new intervention in the mid-1970s. This was feminist geography. Human geography, like other academic fields, had a history of quite appalling sexism, despite the fact of a number of outstanding female geographers, like Alice Garnett, Hilda Ormsby, Eva Taylor, and later Monica Cole, all working in Great Britain. The spatial–quantitative revolution was vigorously masculinist. Women were underrepresented in geography[30] and particularly in spatial–quantitative geography.[31] The atmosphere shared much with that of the economics of the era, an economics that it admired and drew on in its application to location theory: a hard-nosed, cerebral, rational approach to the world appealing to gender stereotypes[32] and sometimes matched by an unpleasant masculinist chauvinism, which did not stop short of plagiarizing the serious contributions of women geographers.[33]

By the early 1970s the women's movement had begun to make its mark, and this had an effect on academia. The number of female graduate students had increased, and there was increasing pressure to hire them. Slowly the gender composition of department faculty started to change, and given the broader social science tenor of the times, this would be reflected in research themes and emphases. Women's geography would henceforth be on the map. This was apparent in a welfare geography more attuned to issues of gender inequality; and in that social reproduction that was a complement to the production emphasized in Marxist geography.

To situate the first interest: An early publishing venture that tried to exploit the socially relevant geography interest more generally was McGraw-Hill's Problems Series in Geography. It included titles like *The Geography of Poverty in the United States* and *The Black Ghetto*, but significantly nothing on gender. By the middle years of the decade, this sort of neglect would no longer have been possible.[34] Female geographers were asserting a distinctly feminist viewpoint that, initially at least, focused on the way women, in virtue of gender roles, were disadvantaged relative to men by geography. A major theme, continuing well into the 1980s was, therefore, that of the spatial entrapment of women; their confinement to the sphere of reproduction, to the home, and how that limited employment possibilities. These arguments entered into early attempts to make sense of the suburbanization of back office employment (Nelson, 1986). In virtue of her dual role, the married, suburban woman had lower wage expectations, could be expected to be less militant than her central city, often divorced, counterpart, and this made her an attractive recruit for the deskilled work of the back office. This is a theme taken up from a more general angle in the work of Hanson and Pratt (1988, 1990, 1991) on gender differences in labor markets. Women seemed to search for jobs closer to their homes and were more likely to obtain information about possibilities from family and friends than were men, though the degree to which women were indeed spatially entrapped would be critically examined later by England (1993).

More closely related to the Marxist work were those feminist studies that, in a sense, tried to make good Marx's overwhelming focus on production by an intensified study of the modalities of social reproduction. Marx had little to say about the reproduction of labor power, which gave rise to the view that he had disparaged this, albeit mediated, crucial contribution of women to production. Someone who has been conspicuous in her investigations of this has been Linda McDowell. She early (1983) recognized the role of the unpaid domestic labor of women in supporting "the structure of capitalist cities" (p. 60)—not least, meals, laundry, free transport, and child care for others—and has always been a strong advocate for bringing studies of production and reproduction together.[35,36]

Subsequently, with Doreen Massey (1984), she explored how this worked out over space. What she and Massey were particularly interested in was showing how gender and capital had articulated in different ways in different parts of England in the late 19th century and how, with changes in the division of labor, including its spatial expression, the relation had changed. In the coal-mining areas of County Durham, the relationship had been very different from that in the cotton textile areas of Lancashire. In the former, there were limited possibilities for wage work for women. Most were confined to the extensive amounts of domestic labor required by coal mining at a time before the introduction of pit baths or domestic bathrooms. The danger of underground work contributed to a male solidarity, which was, in turn, a condition for a close social life outside of work that isolated women still more. In Lancashire opportunities for wage work for women were much more extensive, and this was a basis for unusual levels of unionization among them. Lancashire also became an important base for the development of the suffragette movement lobbying and demonstrating on behalf of votes for women: in other words, a very different sort of outcome than what would have been conceivable in County Durham.

One of the important features of McDowell's work, though, has been the way in which she has drawn attention to social reproduction as not just a material process in the narrow sense but also a cultural one. In a thought-provoking article in 2004, she outlined the gender and class implications of occupational change in the advanced capitalist societies. The context attracting her attention was the growth at the lower end of the occupational scale of what she called "interactive" work: work, as in the hospitality industries, retail clothing, even fast food, where self-presentation, in both emotional and physical terms, is important. As she points out, it is this sort of work, which, in a context of deindustrialization, is increasingly all there is for those who lack credentials or skills. These are jobs requiring docile, deferential performance and are often constructed as feminized. Yet in contrast, a feature of male working-class youth is an exaggerated masculinity. This is interpreted by McDowell as a protest against what she calls "middle class norms of sobriety and conformity"; "high touch" jobs are seen as "an affront to their masculinity" where masculinity is defined as toughness, independence, and sexual aggressiveness.

McDowell does not, however, develop the connections of these observations with traditional themes of inquiry in human geography. In this particular article, she says that place matters, but to the reader it remains unclear how. This is in some contrast to Massey's (1995) innovative work on the workplace/living place geographies of high-tech workers. Far from the rigid separation characteristic of most work, this is one in which a masculine-dominated sphere of work invades the domestic space, while the reverse does not occur. So, on the one hand, family life is disrupted, creating all manner of tensions, yet none of the employers in this particular instance provided crèches. The picture Massey paints is of a group of men so in love with their work, the logic at the heart of it, and a sense of importance deriving from the "scientific" nature of their work, that they have little time to devote to a living place purged of work concerns.

THE ADVENT OF *SOCIETY AND SPACE*

In 1983 a new journal made its appearance: *Society and Space* would be part of the *Environment and Planning* family. Its editor was Michael Dear, and the editorial board consisted of Allen Scott, John Short, and Nigel Thrift. Of a 17-person board, only Simon Duncan and possibly Doreen

Massey could be regarded as bringing strong Marxist sensibilities. Something different yet with continuities with the way human geography had developed during the 1970s was clearly emerging.

One year earlier, in 1982, Duncan and Ley published an article with the provocative title "Structural Marxism and Human Geography: A Critical Assessment." This was a paper that signally failed to come to terms, to even recognize, the dialectical character of Marxism, but it undoubtedly expressed some widely held views of the time. The article's technique was to identify certain one-sided approaches to the world with structural Marxism—the very word "structural" was a clear and carefully chosen provocation—and in opposition to what was being left out. Accordingly, structure gave short shrift to the claims of agency, Marxism privileged wholes over parts, the economy over culture, theory over facts, class over ethnicity, gender and stratum. The facts might square equally well with other theories. In talking about housing markets, Weberian-inspired urban managerialist theory might be just as valid an explanation as Harvey's emphasis on class-monopoly rent. And land use conflict could be better understood in terms of a struggle between interest groups as between capital and labor.

In reflecting on those claims in 1994 (pp. 107–108), Derek Gregory, who had been appointed to the editorial board of *Society and Space* a year after its foundation, thought them misconceived and unfair. That may be. But while far from a manifesto for "society and space," the article did contain much that would resonate among its protagonists. While not *anti*-Marxist in the way that the Duncan–Ley article can claim to have been, despite its attempt to hide behind different sorts of Marxisms with human geography having opted for the "wrong" one, the interest in "society and space" did seek some critical distance from it. Historical materialism had been *the* critical theory of the 1970s. Nothing else was on offer. For a variety of reasons this was believed to be unhealthy. Theory was too important in any science, the human sciences included. And in a number of respects historical materialism did not meet the challenge. In part, this reflected some of the issues that Duncan and Ley had touched on, though suitably reworked and thought through.

Likewise, there was real concern in the human sciences about the need to come to terms with the dualism of structure and agency. Interpretive social science emphasized the latter: how people imposed meanings on the world, drew upon a shared body of meaning, and acted in terms of those meanings to create new structures of social relations, which would then be related to but always in terms of ever shifting interpretations. Marxism was seen as overemphasizing the role of structure in the understanding of social action: above all the limits placed upon action by the necessity to accumulate, the necessity to speed up the circulation of capital and the like.

There was also sympathy for the critical view that historical material-ism was economistic and downplayed other dimensions of human experi-ence. Feminist geography was more than a straw in the wind here, imply-ing, in its reference to gender roles, that there might be other structures of relations than those of capital and at the same level of abstraction. Equally if not more significant in giving some theoretical underpinning to this was the interest that geographers showed toward the end of the 1970s in critical realism. Critical realism was a philosophy of science but through its methodological procedures it implied an ontology of separate structures of social relations: structures like the division of labor (Sayer and Walker, 1992), capital, gender (Foord and Gregson, 1992), the state (Lovering, 1987), and so forth. The methodological procedure at issue was its predilection for what Gunn (1989) has described as "empiricist abstraction." This was not the Marxist approach to the question, though, and more is said about this in Chapter 6.

Finally, there was indeed the whole question of space. Space had been the all-consuming passion of the spatial–quantitative revolution. Bunge's wonderful book, *Theoretical Geography*, had shown just how far a conception of space as relative could be pushed, and it was very persua-sive.[37] This was not least because the emphasis on geometry readily elic-ited responses of an aesthetic sort. More fundamentally, it was because of its ability to disclose aspects of the spatial organization of society that had, hitherto, been hidden, buried in a mass of detail waiting to be sorted out by the spatial analyst. With the turn to Marxist geography and social relevance, something seemed to have been lost. The spatial–quantitative revolution had been an approach in which spatial understandings did indeed seem to have a cumulative character. Central place theory could be examined and reexamined depending on historical context, on level of economic development and so forth, and seemingly expand knowl-edge that was undeniably of a geographical sort. Now, in socially relevant geography, space seemed to lose the preeminence it had enjoyed. Where, for example, were the spatial regularities in welfare geographies or in the geography of modernization, other than those between centers and peripheries or between quite concrete geographic categories like central city and suburb? And once identified, what could one *do* with a distinc-tion between center and periphery? Marxist geography also struggled for answers. Social relations clearly conditioned geographies in a very general sense: With capitalism living places were separated from workplaces, with the result that politics too bifurcated into a politics of the living place and one of the workplace. Likewise, cities bore a patriarchal imprint.[38] But how might space condition social relations, and how could the two sides be put together in some sort of relationship that was nondeterministic?[39] Only with the appearance of Harvey's *Limits to Capital* (1982) did things become clearer, but by then the interest in something rather different, "society and space," was already well underway.

Centrally involved in this were Derek Gregory, Nigel Thrift, and, to a somewhat lesser degree, Andrew Sayer. And if one person above all made a formative impression on their thinking during the late 1970s, it was Anthony Giddens.[40] Giddens was a British sociologist who was extraordinarily prolific and who acquired a responsive audience not just in sociology but in geography too. Part of the attraction for geographers was the fact that through what he called "structuration theory" he tried to theorize society and space together and to draw out the implications of such thinking for our understanding of the world. For Giddens social relations were spatial relations and space was implicated in their very nature. An absolutely central idea was that of time–space distanciation. Through the technologies first of writing and then of, successively, the stagecoach, the railway, the telegraph, and the telephone social relations had acquired increasing depth and breadth: "depth" in the sense that relations could be established with people in the past through records about them or their writings, and breadth through the ability to span large slices of space and elude the requirements of the face-to-face. Money was crucial to modern processes of time–space distanciation in the way it allowed a bracketing of spatial and temporal barriers and mediated between those who might never come face to face.

Time–space distanciation was implicated in power through the idea of the storage of resources: resources were centralized and then drawn on in further rounds of distanciation. What he called "authoritative resources," including information, were centralized and then used in empowering ways through the surveillance that they enabled. "Allocative resources" were of a material sort stored in the form of, for example, fixed capital or livestock; creating a surplus presupposed the existence of storage devices. And as a final example, Giddens drew attention to the implications of time–space distanciation for the legitimation of the social order. The face-to-face was the sphere of social integration and the means through which normative commitment was secured. The interplay of presence and absence implied by time–space distanciation, though, meant that securing that commitment was put at risk as those to be integrated might not be so available at all times.

Giddens's emphasis on the spatial and particularly on the articulation of space and time was undoubtedly a major reason for his attraction to human geographers. There were, however, others that made him appealing to those human geographers looking around for some approach that would respond to what they saw as some of the weaknesses of historical materialism. In particular, through structuration theory, he seemed to provide an answer to the problem of agency and structure that promised to replace a dualism with a recognition of the way they were co-constituted. Very simply, agents were always agents in virtue of a social structure; social structure mediated their actions. It limited what they could do, prescribed what they had to do while, at the same time, enabling in certain ways. Agents were always socially formed, in other words. People acted according to prescribed roles, and could also change those roles but only by drawing on other structures of social relations. On the other hand, through their actions people reproduced those social structures and might even transform them. To the extent that people acted in accordance with the rules laid down by a social order, then it would be reproduced.

Second, Giddens promised an escape from the totalizing vision of historical materialism. The very idea of time–space distanciation and the way the interplay of presence and absence allowed some discrepancy between social order and normative commitment suggested that societies were not the integrated structures that theories of totalization implied.[41] His concept of power was also significant. This was because, as we have seen, he postulated the existence of two sorts of power—allocative and authoritative—implying a separation of state and capital that Marxist critics were quick to seize on.[42]

Third, there was a rejection of the evolutionism that many attributed to Marx: the notion that social change had a direction. In some of Marx's writings, particularly, the Preface to *The Critique of Political Economy*, this was very clear: the claim, that is, that social relations changed in such a way that the productive forces could continue to develop. This change, moreover, was necessitated by the way in which the productive forces encountered barriers in the existing social relations: contradiction, in other words, that would be suspended without radical change in the relations of production and then abolished as a new mode of production came into being. However, for Giddens, and consistent with his view on tendencies toward time–space distanciation and hence the inherent disunity of social relations, the very idea of society had to be in question. That social relations changed, there was no doubt, but it occurred along what he called time–space edges where different social forms came into contact with one another facilitating transition to something new. The idea of societies proceeding through different stages was, therefore, rejected, and rejected by an appeal to notions of spatial difference about which, for the most part, Marx had been silent.

Fourth, and finally, Giddens's arguments coincided in some respects with those of the postmodern thinkers with whom a number of those associated with the journal *Society and Space* were increasingly sympathetic. As Nigel Thrift, one of the *Society and Space* troika, put it in the debate with Harvey in 1987 and in response to him:

Recently this kind of discourse has been making many commentators more and more uneasy, both within and without Marxism. First of all, societies seem to be only partially integrated. They seem to be predicated not just on the universalizing tendencies of various economic, social and cultural processes but also on the genuine differences that emerge as these processes are worked on by all the different people that inhabit all of the different spaces from which they come and in which they come together. The result of this work can be spaces in which new institutions and interpretations can grow into new processes. (p. 404)

I say more about this in the chapter to follow. Meanwhile, I should note the quite extraordinary influence that Giddens had on human geographers of a more critical persuasion, and particularly those who thought that Marxism was not the answer to human geography's problems. A book published in 1985 and edited by Derek Gregory and John Urry brought together some Marxists, notably Harvey and Dick Walker, with a number of people from the "society and space" stable. Of the 15 chapters in the book, 10 make reference to his work (excluding Giddens himself in his chapter) for a total of 32 references altogether and Giddens is the most cited of anyone; Gregory alone made 11 of them. This contrasts with references to Marx, whose works are cited a total of 28 times in 7 of the 15 chapters; but 12 of these are in Harvey's chapter and 10 in Walker's.

In short, much had happened between Harvey's publication of *Social Justice and the City* in 1973, the book in which he first laid out his vision for a Marxist geography, and the appearance of the new journal *Society and Space* a decade later. New faces had appeared along with new ideas. Structuration theory would eventually be seen to have promised more than it could deliver, but at least it stimulated theoretical debate where it had previously been lacking. The importance of theory was reaffirmed. Something else had happened. Marxist geography had generated a sense of political commitment. The cover of its major journal *Antipode* bears eloquent testimony to that: humanity breaking free of its chains of oppression (Figure 3.1). The cover of *Society and Space* is very different but equally eloquent. The politics seems to have disappeared. Instead, we have people, individuals for the most part, going about their daily business, shopping, walking in what might be a park, lovers holding hands (Figure 3.2). To a degree this is deceptive. *Society and Space* would foreground theory and theory is inevitably political. This would be very apparent in the articles it published. It would, however, be a different sort of politics: more ameliorant and less confrontational, less violent in its implications, more partial and less total, and so lacking some clear social vision. In that regard and in numerous others, "society and space" would be a bridgehead to that further development of the encounter with social theory occurring under

THE SPATIAL FIX – HEGEL, VON THUNEN AND MARX

MARX ON HISTORICAL FORMS OF THE PROPERTY RELATION

LANDSCAPES, SOCIAL RELATIONS AND ECO-GEOGRAPHY

URBANIZATION IN SOVIET SCHOLARSHIP

HOUSING AND PROLETARIAN LIFE IN EARLY CAPITALISM

BIRDS IN EGG/EGGS IN BIRD

VOLUME 13 NUMBER 3 1981

FIGURE 3.1. Cover of *Antipode*. Copyright 1981 by John Wiley & Sons. Reprinted by permission.

the auspices of the "posts." It is to those changes that I turn in the next chapter.

NOTES

1. "The quantitative movement can thus be interpreted partly . . . as a response to outside pressures to discover the means for manipulation and control in what may broadly be defined as 'the planning field'" (1973, p. 124).
2. Consider Harvey's (1973, p. 95) comments on how to deal with the urban question just prior to his embrace of Marxism: "In part it is a problem of exercising a wise control over social and spatial organization within the city system. Here an enormous task confronts us. We really do not have the kind of understanding of the total city system to be able to make wise policy decisions, even when motivated by the highest social objectives. The successful formation of adequate policies and the forecasting of their implications is

FIGURE 3.2. Cover of *Society and Space*. Copyright 1992 by Pion Ltd. Reprinted by permission.

going to depend on some broad interdisciplinary attack upon the social process and spatial form aspects of the city system."

3. Compare Jacques Lévy (Lévy and Lussault, 2003, p. 985) in a discussion of French geography: "Sous l'influence de Vidal, la géographie française . . . se situait à l'intérieur du courant de la totalité organique en posant comme postulat l'intégration principale de l'homme et de son environnement au sein d'un ensemble ayant sa propre cohérence qu'il convient de mettre en jour. Cette géographie fût, sans surprise, peu encliné à prendre en compte des logiques proprement sociales, telles que l'urbain ou le politique." ["Under the influence of Vidal, French geography . . . situated itself with respect to the idea of the organic totality. It did this by emphasizing the integration of man and his environment at the heart of an ensemble whose coherence it aimed to investigate. Unsurprisingly, this approach was not especially inclined to take into account social logics like those of the urban or the political"]. The argument applies equally to anglophone geography.

4. Today he is best known for his time geography, but the earlier work on migration and the diffusion of innovation was the more startling in its implications. Hägerstrand was a master of the simple but powerful idea.

5. He thought of it in terms of an historical process. Quite by chance people might have found themselves at a particular destination. For some, again chance, reason they stayed there. They retained some contact with the people back home and provided "information" about jobs and housing there. Some of these might then follow, and in their turn provide the same mediating function for yet other waves of migrants from the same place in a positive feedback process. This is the process now known as chain migration, but Hägerstrand was the one who first theorized it in terms of some spatially structured flow of information.

6. Certain social regularities in voting behavior were well known. Blue-collar workers tended to vote for left-wing parties and white-collar workers for parties of the right. But there were also some spatial regularities. Blue-collar workers living in blue-collar neighborhoods were even more likely to vote for Socialist or Communist parties. Could it be that, again, there was some sort of distance-biased communication structure? Was it that people's party political identifications were formed within some sort of local climate of opinion and sentiment communicated within the locality in some sort of neighborhood effect analogous to Hägerstrand's diffusion of innovation process? This was to play some part in the resurrection of a political geography that, in Brian Berry's famous words, had become a "moribund backwater."

7. Brian Robson's (1969) innovative work on the role of social networks in accounting for variation in educational attitudes is also notable. The characteristics of particular neighborhoods certainly counted for something: Less affluent people in materially deprived areas tended to have less positive attitudes; but if their social connections were outside the neighborhood this would change and attitudes would shift to the more positive end of the spectrum.

8. See Derek Gregory's (1985) excellent critique.

9. Reference should also be made to William Bunge's *Detroit Expedition*. Like numerous others at that time who had been in the vanguard of the spatial–quantitative work, Bunge now made a radical change of direction. Based at Wayne State University in Detroit, he conducted a number of field research exercises into the social and political geography of the city, often in highly original ways both in mode of presentation (a map of infant mortality rates by subarea in which each subarea is given as a label the country, often a very poor country, with a similar rate) and what was being presented (e.g., the distribution of rat bites of children, the distribution of auto accidents involving children). A notable outcome of his work was the book *Fitzgerald: Geography of a Revolution*.

10. In a limited sense Wolpert's work on locational conflict had been anticipated some 5 years before by Harold Mayer's (1964) study of conflict around the future of the Lake Michigan shoreline southeast of Chicago. It had none of the effect of Wolpert's work, however, suggesting that theoretical sensibilities had already been stirred among human geographers and Mayer's failure to theoretically situate his work meant that it would not have the impact it might otherwise have had. One can make the same point about the article by Spencer and Horvath (1963) on the origin of agricultural regions; it was surely an exemplar of the "new regional geography" (see Chapter 5) without knowing it and remained unable to situate itself theoretically.

11. Given the supradisciplinary character of Marxism, to talk about Marxist geography is, of course, oxymoronic. Yet we continue to live in a world of disciplines and to refer to Marxist economics/sociology/political science/ anthropology. Marxist geography also remains distinctive in the sense of a Marxism that emphasizes the spatial moment of the social process. It will be retained in this book.

12. "Liberal" in the sense of progressive and left leaning.

13. The major figures in this development were Gilbert White at the University of Chicago and his former students Robert Kates and Ian Burton.

14. Compare Robbins (2004, p. 52): "Many and perhaps most practicing political ecologists would in no way identify themselves as Marxists or materialists."

15. As Wylie (2007, p. 23) has commented, "In some respects, therefore, this was a cultural landscape geography more at home in the country and the past than the city and the present."

16. In British geography, landscape studies were quite different. Landscapes were to be seen as historical artifacts, the results of layerings of various humanizations, as Darby (1951) put it and without moral connotation or necessary visual appeal. Landscapes were records of human work on the land: in Darby's (1951) case, the clearing of the woodland, the draining of the marshes, the reclamation of wasteland, the creation of industrial towns, and of country seats. Landscape study involved a peeling away of these different influences.

17. On landscape as national icon, see David Lowenthal (1991).

18. His article targeted in particular the work of Paul Sweezy, Andre Gunder Frank, and Immanuel Wallerstein.

19. See Corbridge (1986) for a critique from within human geography that draws on Brenner.

20. Another instance is Knox, Agnew, and McCarthy's *Geography of the World Economy* (2008).

21. By this I don't mean "difficult" in the sense of some branches of mathematics. The problem, rather, was one of living in a capitalist society, assimilating from an early age the categories through which one could live in that society and navigate it as well as being socialized into a, usually conformist, view of how one fitted into it and then being asked to look at the world through very, very different spectacles: spectacles that allowed one to see different things and connections but which then, if sense was to be made of them, had to be translated back into the categories of bourgeois society.

22. Republished as Chapter Two in *Social Justice and the City* (1973).

23. On the other hand, there is clearly a demand for this sort of work that extends beyond professional geography readerships, particularly when it is in atlas form. The work of Danny Dorling (1995; Dorling and Thomas, 2011) is an excellent example. One is also reminded of David Smith's (1979) evocatively titled *Where the Grass Is Greener: Living in an Unequal World*; while not an atlas in the same sense, even though it included numerous maps, it was an attempt to appeal to a wider audience.

24. I use the term "utility maximizing" both deliberately and accurately since the concept of markets drawn on by socially concerned geographers typically mimicked that of neoclassical economics.

25. Wolpert's influence was especially clear here. The work of some of his students, particularly that of Michael Dear and Jennifer Wolch was of this genre.

It is also very apparent in the collection of articles that I edited with Ron Johnston in 1982, even while this particular approach was by then undergoing some eclipse. The titles of the three major sections of the book are indicative: Part One: Locational Conflict; Part Two: Institutional Context and Conflict over Location; Part Three: Beyond Locational Conflict.

26. See the discussion in Chapter 4, particularly of David Ley's empirical work.

27. See Scott's (2000) excellent outline of this history.

28. As an implicit critique of the orthodoxy, it would be stated most clearly in the distinction that Storper and Walker (1989) drew between weak and strong competition. Restructuring clearly falls into the category of the latter while weak competition has always been the assumption in neoclassical economics.

29. See also Clark's (1993) critique of the new industrial district literature.

30. I started graduate school at the University of Illinois in 1961. There must have been at least 20 of us. There was only one woman, and she was the first that the Department of Geography there had ever accepted for graduate study. Janice Monk (2004, pp. 10–12) has an interesting story to tell about her experience as well as about other female graduate students at that time.

31. There were some interesting exceptions. Duane Marble always encouraged female graduate students, and they included ones like Sophie Bowlby and Susan Hanson who would go on to have careers of distinction.

32. Even as recently as 2004 Linda McDowell could write: "Masculinity and femininity are defined in relational terms; masculinity is what femininity is not: not emotional, not empathetic, not muddled or messy, not constituted in relations of dependence but instead cool, rational, cerebral, strong and independent" (p. 45).

33. There is some evidence that Fred Schaefer's famous "Exceptionalism in Geography" was not entirely his work; that portions were filched from the work of a female graduate student. See Corry and Sugiura (1999).

34. However, of the two collections coming out of the (British) localities project (Cooke, 1989; Harloe et al., 1990), only in the first volume were any of the contributors women and then only two out of 28. And only in the Cooke volume was any attention paid to gender issues (two of the seven locality-specific chapters).

35. "The focus of Marxist and feminist urban studies must be the interrelationship of production and reproduction *as part of a single process*" (1983, p. 62). See also her vigorous riposte (1986) to a critical realism-inspired attempt to claim patriarchy as a structure of social relations separate from that of class.

36. A further effect of this initiative was to draw attention to the significance of the sphere of social reproduction and its geographies. Cindi Katz has referred to the effects of globalization on the spatial displacement of reproduction functions through migrant labor and nanny chains, for example (Katz, 2001).

37. It is still worth reading. For all its shortcomings, I would regard it as one of the major books in human geography in the 20th century, not just for its missionary zeal, and there was plenty of that, but because of the sheer imagination shown in teasing out spatial relations and relating them to an economy of time.

38. See Massey's (1991, p. 47) discussion of this.

39. Compare Massey (1985, p. 12): "The focus of the radical critique of the 1970s was far less on the huge variety of outcomes we see in the world around

us than in unraveling their common underlying cause. The argument on the whole was not just that spatial patterns are caused by social processes, but that they are caused by *common* social processes. . . . My summary critique of the 1970s would be, then, that "geography" was underestimated; it was underestimated as distance, and it was underestimated in terms of local variation and uniqueness. Space *is* a social construct—yes. But social relations are also constructed over space, and that makes a difference." Soja (1985, pp. 111–112) came to a similar conclusion: "Throughout the 1970s, it appeared that an explicitly materialist interpretation of spatiality would remain lost in an ambivalent dance-step which moved forward only to glide back again, a space versus class *gavotte* choreographed to avoid the long-established dangers of diversionary and bourgeois spatialism. So seemingly equivocal and inconclusive was this dance that it provoked observers (as well as some participants) to conclude that spatiality could fit into Marxism only as an ancillary detail, a perfunctory reflection of more fundamental social relations of production and spatially abstracted 'laws of motion.'"

40. Not uncoincidentally, Giddens would be one of the original members of the editorial advisory board of *Society and Space*.
41. Though this rested on a widespread misunderstanding of what totalization means, something I return to in a later chapter.
42. For example, E. Olin Wright (1983).

CHAPTER 4

Social Theory and Human Geography: Worlds of Meaning

The critical reaction to the spatial–quantitative revolution in the 1970s had been a strong one. The social theory at the core of location theory had been discarded in favor of approaches that foregrounded power and the political. There had been theory in the form of the Marxist work but also more eclectic understandings that, although useful in shedding light on the empirical, lacked the former's coherence and explanatory penetration. What held them together, though, and for much the larger part, was an emphasis on material relations: on the labor process, on markets, on political interventions. From the early 1980s on, this predominance would begin to wane. It did not happen very rapidly but the signs were already there, and prior to the appearance of Soja's *Postmodern Geographies* in 1989, which would lend something of an imprimatur to the turn human geography was taking. For instead of the material, what was now being emphasized was the significance of meaning, the communication of meaning, and the ways in which meaning conditioned the production of space. This would eventually metamorphose into an embrace of the so-called "posts": postmodernism, poststructuralism, and postcolonialism.

The new interest was apparent in the journal *Society and Space* from the start. Michael Dear's editorial introduction to the first issue made reference to Foucault. The trend was clearer 4 years later in a special issue in which a number of people, interestingly all male, debated Harvey's historical geographical materialism. The anxieties about metanarratives which the postmodern turn would entail are already apparent there in Michael Storper's (1987) "The Post-Enlightenment Challenge to Marxist

Urban Studies," where he is keen to underline the way in which Marxist theory, like any theory, excluded certain questions because of its own priorities. The tendency was also clear in Nigel Thrift's (1987) vigorous rebuttal of Harvey's claims about totalization and the possibility of any privileged viewpoint.[1]

The new understandings would not be reducible to the question of meaning. Far more was at stake than that. Rather, particularity and difference would assume a new significance, as would the relation between power and discourse. New conceptions of space surged to the fore. Context would be emphasized as the homogenizing effects of capital were subject to a new skepticism. The interest in power was still there but now hitched to the problematic of discourse rather than to the world of production and exchange. Theory received a new boost, but it was a very different theory from that embraced by the Marxists.

Some of this had been anticipated in the 1970s in the form of what had been known as humanistic geography. This too had lined up against Marxist claims, but it lacked the commitment to theory that the "posts" would provide. Nevertheless, it was an important precursor, and there would be continuities. It is with humanistic geography, therefore, that this chapter begins. This is followed by a discussion of what, for shorthand, I call the "posts" and their implications for geographic understanding. The chapter concludes with some summary comments on the meaning of the engagement of human geographers with social theory *toute entière*. Here I focus first on what it has meant for understanding in human geography and then the implications of how it has unfolded for contemporary research emphases.

HUMANISTIC GEOGRAPHY

The contexts of that rediscovery [of humanist principles and aims] . . . were initially negative or, at least, broadly critical. For all its munificence, the overarching growth ethic, highlighted by an explosive technology both on the planet and in extraterrestrial space, increasingly appeared as the bearer of monumental self-destruction. The convergence of science and technology, once the Promethean harbinger of utopian society, began to emerge more as a central villain in the exhaustion and despoliation of man's own environment. The linking of scientific rationality and politics, once the hallmark of enlightened democracy, moreover, began to emerge as the chief mechanism for a stronger, if more subtle and therefore less penetrable, despotism. . . .

That the sciences, and especially the social sciences, either reinforced the status quo or else ignored the ethical debate

surrounding the intellectual community in the name of neutrality,
detachment or objectivity, seemed, at least to many, an admission
of academic irrelevance. . . .

Protest was directed not only against the *actions* of a technological
society but its *cognitive categories*, its alienating worldview, built
around the mystical glorification of technique.
—LEY AND SAMUELS (1978, pp. 1, 2)

Marxism was not the only standpoint from which an external critique of
spatial–quantitative geography was launched. With some delay a very different critical window was opened. The focus here was less that of what
had become known as social relevance and more that of the technocratic
climate within which the spatial–quantitative revolution had found such
fertile soil. Certainly this resonated with the sorts of planning issues that
had become increasingly politicized during the 1960s: ones having to do
with the housing dislocations caused by urban renewal and the construction of freeways. But as the prior quotations demonstrate, this alternative
critique was in no way materially based as that of Marxism had been.
It wasn't the contradictions of capitalist development that were now at
issue. Rather, it was a matter of ideas: ideas that were in their application
oppressive across all dimensions of social life. Accordingly, and as itself
an expression of those dominant understandings, the spatial–quantitative revolution had to be much more centrally in question than had been
the case with the Marxist critique. In short, it seemed that the cultural
and historical geographers that had been so discomfited by the spatial–
quantitative revolution might have finally found a banner behind which
to march.[2] Significantly, though, for virtually all the disaffected spatial–
quantitative geographers it was to be historical materialism, the new critical economic geography, or welfare geography that would be the new
poles of attraction, and not humanistic geography.[3]

Humanistic geography's take on the events of the 1950s and 1960s
was very different—indeed, a take so different one might wonder what
it is doing in a chapter on social theory and geography. For rather than
putting social relations at the center, it was the individual that was to
take pride of place. Humanistic geography celebrated the individual, her
complexity, her irreducibility. Emphasis was placed on human experience
and the meanings that people ascribed to those experiences, including
their emotional responses, the values that they brought to them, and the
considered actions informed by those meanings and values. Nevertheless,
in the course of time, the contribution would be an important one in the
encounter between human geography and social theory.

We can grasp some sense of its self-understanding through a series of
oppositions (see Figure 4.1). In the figure, the terms on the left indicate

| Subjective–Objective |
| Concrete–Abstract |
| Place–Space |
| Particular–Universal |
| Contextualizing–Generalizing |
| Understanding–Explanation |
| Action–Behavior |
| Meanings–Facts |
| Interpretation–Cause and effect |
| Ideal–Material |
| Local knowledge–Theory |

FIGURE 4.1. Humanistic geography's dualisms.

the emphasis of humanistic geography, at least as it saw itself. On the right, I have identified what were, in effect, the *bêtes noires* for each. It associated these not just with the spatial–quantitative revolution but also with its Marxist geography rival. In humanistic geography, it was argued, people *acted*. They were endowed with intentionality and an ability to construe courses of action. They acted as concrete people, drawing on the full diversity of their experience. This was in contrast, or so it was claimed, to the "pallid, entrepreneurial figures" (Ley, 1977, p. 499) of Marxist geography and location theory. These were treated as abstractions of their true concrete nature. They behaved. They blindly conformed to the dictates of the space economy and its price and wage gradients. Capitalists were just that and no more.

The understanding of meaning was central to the task of the humanistic geographer and would have important reverberations for the development of human geography. Just what did particular actions mean to those carrying them out? The actions expressed an understanding of the world, both in its cognitive and normative dimensions, but exactly what were those understandings?

Through their actions, people constructed their worlds. Behind the objective, Ley (1977) argued, was always the subjective. The abstractions of the social scientist could never grasp that world since it was the concrete person in all her complexity who acted. And even if abstractions could be brought to the point at which the concrete could be more adequately grasped, that would still not suffice since it could not come to terms with the sheer creativity and therefore unpredictability of people. The most that could be hoped for was to bring meaning-imputing agents into a relationship with the particular environment or context with which they interacted and so had to construct meanings. Meanings were saturated with the concrete: the concrete nature of the person and the concrete nature of the context with which he or she interacted. People didn't interact with a context conceived in the thinly abstract terms of geometry. Rather, it was the sheer concreteness of particular ensembles of things, including people, that were at issue.

All this meant a reordering of the focus of geographic inquiry. It could not be space as the advocates of spatial–quantitative geography had argued and as was implicit in the work of Marxist geographers with their talk of geographically uneven development and logics of urbanization. Rather, geography had to concentrate on place. Place was concrete: It meant different things to different people, depending on the experiences and values that they brought to it. It was, therefore, subjective. Space, on the other hand, was conceived as something abstract and determinant, bearing down on individuals, determining their actions, and abstracting from the specificities of place. If human beings were as humanistic geographers assumed them to be, utterly and irreducibly individual, then this was an impossible project.

Even so, humanistic geography was more complex and diverse in its practice than this brief overview might suggest. Partly this was because it wasn't working on an entirely blank sheet. There had been antecedents, and once a self-conscious humanistic geography made its appearance in the 1970s, they didn't go away. To some degree the different tendencies kept their distance; to some degree there was an element of rapprochement. Central themes had already been outlined in the 1960s and in the absence of serious anxieties about the spatial–quantitative work or the planning ethos for which it would come to be seen as a response. David Lowenthal and Yi-Fu Tuan were important precursors and would continue alongside what would emerge in the 1970s as a more analytic strand of humanistic geography.

Of the two, Lowenthal was the less theoretically engaged, and I express it thus because the role of theory in humanistic geography was to become an issue. Lowenthal's sources were and continue to be those of the humanities: largely literature, history and philosophy—particularly literature—and these have been the basis for his attempt to identify universals in the human condition that might shed light on our relation to places: how, for example, landscapes are always seen through the prism of our memories (1975); how literature influences our responses to them. A paper on islands (1991) then explores what he calls the insular condition, again as understood largely through a welter of meanings in literature and testifying to a quite extraordinary intellectual reach. The search for universals about people and how they relate to place and landscape is the fundamental theme in his work.

This was a continuation of work that Lowenthal had started in the very early 1960s (e.g., 1961). Tuan started a little later but was certainly engaged in a recognizably humanistic geography before people came to use the term. His paper of 1971 is a case in point. It is perhaps the first to draw on the phenomenology of Edward Husserl and his search for essences: the attempt, that is, to isolate the essential aspects of objects from all preconceptions, whether scientific or otherwise. In Tuan's hands, this became a search for the essences of people as essentially geographic beings. He discusses the human propensity to seek meaning in order or harmony and not necessarily for practical reasons. We all live in centered worlds of our own but can join with others so as to give meaning a mark-edly ethnocentric character. Part of the desire for symmetry is expressed through thinking in terms of binaries: home acquires its meaning in the context of journeys—home as refuge, as certainty. Essentially place is a feeling: a very different view, therefore, from that embodied in the regionalization algorithms of spatial–quantitative geography.

Ley's humanistic geography, the humanistic geography which called itself thus, which emerged in the 1970s in reaction to the spatial–quan-titative work, and for which he was the most prominent advocate, was quite different and, in its convergence on social geography, substantially more eclectic—and this despite the fact that Ley was indeed a persuasive advocate for a human geography that, like that of Lowenthal and Tuan, put individual experience, meaning, and values at its center. His article (1980) on the politics of the postindustrial city provides a useful instance.

The first thing to note there is the contemporary, highly concrete, character of the object of study: not much sign of a concern for essences there. Second, there is an engagement with power relations that again seems quite alien to the humanistic geography of people like Lowenthal, Meinig, or Tuan and suggesting a more finely developed sense of the importance of the social. What the paper focuses on is a struggle around policy in the city of Vancouver between those espousing very different values. A professional-technical-managerial stratum had emerged with the postindustrial city of the 1960s and 1970s and opposed, ironically enough, a rational bureaucratic worldview centered on growth and work. There was a growing concern for what Ley called "values," especially those surrounding conservation, housing densities, the urban landscape. People placed a higher premium on self-fulfillment and lifestyle. On the other hand, there were indeed those allied to a local growth coalition who looked positively to what social engineering could do for the city and regardless of its implications for the built environment.

A second exemplar of this sort of humanistic geography is John West-ern's (1978) chapter on forced relocation under apartheid in Cape Town.[4] Apartheid governments aimed to bring about a rigorous residential sepa-ration of the races. This often resulted in forced relocation, largely of Africans, Indians, and Coloreds or people of mixed race. Western focuses on a particular instance of this in Cape Town where large numbers of

Colored people were removed from an area close to the downtown and relocated at a very, very considerable remove. The reasons for forced removal are discussed, including the economic gains to whites from an ability to gentrify some of the houses from which Coloreds were relocated and the increased spatial security that whites supposedly achieved as a result of the removal of Colored neighbors. There is particular focus on what the author calls "the social costs of forced removal." The elimination of a neighborhood in which people had enjoyed close personal ties, often of a kinship sort, and over many years was utterly devastating for them and mortality spiked. It had repercussions for personal security. This was because in the new residential areas, the absence of social ties meant the elimination of the informal surveillance mechanisms that had held delinquency in check before. Previously, families had watched over the street behavior of the children of other families in a reciprocating manner, but that was no longer possible. There is also discussion of the status differences that emerged among Coloreds subsequent to their relocation and separation into areas of privately and publicly owned housing respectively. Geographically this was expressed in a stigmatization of one residential area by those living in the other.

Before proceeding to a critical discussion of this approach—and there are strong convergences between the two papers, not least the interest in social engineering and its implications—note should be taken of a paper that Donald Meinig published in 1983, which casts it into a quite different relief. His comment on what had come to be accepted as humanistic geography, and including the work of Lowenthal and Tuan, was damning: "Almost all of this avowedly humanistic geography is analytical in intent, aimed at a systematic examination of subjective meaning and human behavior. Most of it seems little more than an extension of science" (1983, p. 315). Meinig, rather, wanted to draw a very sharp line between the humanities and the human sciences and see humanistic geography as drawing its primary inspiration from the former, though as Derek Gregory (1994, pp. 81–84) has argued, a very particular understanding of them. The implication is that it should, therefore, distance itself from Husserl's essences and the social geography of people like Ley and Western. He draws on the use humanistic geography had made of creative literature to develop his point. Humanistic geography had used creative literature as one form of evidence among many. Rather, Meinig felt, it should be contributing to that literature, drawing on the knowledge that is the human geographer's specialty, on an immersion in a particular place and landscape and energized by an emotional insertion in a place to provide evocative constructions of it in a way that would affect the responses of others, including those engaged in literary studies themselves.

What summary comments can one make, therefore? There are clearly some sharp divisions here. First, the humanistic geography of the 1970s was much more clearly rooted in critique: critique of both the assumptions—and *pre*sumptions—of the spatial–quantitative revolution and of the society whose values it seemed to embrace.[5] In turn, this entailed research of a very concrete, though not atheoretical, character: Ley's postindustrial city, Western's apartheid city, or Relph's (1976) placelessness. This contrasts substantially with the emphasis of a longer humanistic tradition in human geography, going back to John K. Wright (1947) and taking in Lowenthal and Tuan, on the universals of the human condition and what they mean for how people experience, understand, and respond to place and space.

Second, Gregory (1994) has detected in much humanistic writing in geography a suspicion of theory. The theory that he has in mind here is clearly social theory. A crucial point that he makes is that conceiving the human condition outside of a world of social relations is incoherent. As he argues, the experience and attribution of meaning so central to the humanistic credo are impossible outside of language, and language is not just a social medium; it is in its origins and transformation, irreducibly social in character.[6] Responses to landscape, and *pace* Meinig, cannot be understood outside of linguistic categories, and these categories themselves reflect social relations and, what is more, the social relations of a particular form of society, which means that the search for essences has to be a qualified one.

In this regard, the humanistic geography that emerged in reaction to the spatial–quantitative revolution was an advance. It was well aware of the significance of the social. It was also theoretical even if the theories it drew on, like that of postindustrial society, had a distinctly middle-range character. And it wanted to come to conclusions about the concrete of the here-and-now. This, in turn, meant that it had to be alert to methodological issues, which, as evident in Chapter 6, it was, not least in drawing on the theory of hermeneutics in order to make critical sense of the relation between the researcher and the people who one wanted to understand. In all these respects, it was very much in tune with that human geography that had emerged along with the spatial–quantitative revolution—one intensely attuned to theoretical and methodological debate.

Even so, while in contrast to other strands of humanistic geography it recognized the significance of the social, and also theory, it had no adequate theory of the social itself other than the fact that people were, well, social: that meanings were intersubjective and those shared meanings constituted something called a life-world. This meant that humanistic geography was ill placed to respond to a materialist critique. Before humanistic geography could be reconstructed along these lines,[7] however, there were further irruptions from outside the field, irruptions that were to provide a much more serious challenge to the claims of historical geographical materialism than it had been able to offer. These were the various "posts," which became part of the theoretical landscape of human geography from the late 1980s on and are still with us: postmodernism, poststructuralism, and postcolonialism. These have sigificantly altered the ways in which human geographers think. They vary in their emphases and to some degree in their understandings. Nevertheless, there have been some shared assumptions.

THE "POSTS"

The Arguments

Significantly, the "modern" or "modernity" to which "postmodernism" refers is that of a particular discursive understanding of the world that emerged in the course of the Enlightenment in the second half of the 18th century in western Europe. I say "significantly" in order to underscore the emphasis the "posts" place on the world of ideas, on what has been thought, how people think, and how what they think is always discursively mediated. This idealistic stance puts it firmly in opposition to the materialism of historical geographical materialism or even of the liberal formulations of the critical mainstream in the 1970s (see Table 4.1).

This "modern" understanding assumed that the world was orderly and that it could be understood through the application of reason. Fortified by the successes of the physical sciences the belief was that just as the physical world could be an object of intervention and so turned to the benefit of humanity, so too could the social world and with the same purpose in mind. Through observation, the organization of observations into higher order concepts, and the development and testing of theory the truth of the social world could be arrived at. On that basis, laws could be developed that would redound to the benefit of all. Through the application of reason and logic, human progress was not only possible, it was inevitable. Two utterly devastating world wars later, with huge loss of life, a world in which the vast majority remains subject to severe material privation, a world in which cooperation around joint projects, like doing something about global warming seems a mere chimera, this vision of progress now seems very tarnished, to say the least. But the impulse of the "posts" is clearly not just to decry; rather, it is to explain why things went wrong and why the modernists could never have been right from the very start: why the world cannot possibly be like what they assumed it was.

The object of modernist thought is modernity itself. Modernity is experienced as something dynamic, continually changing, new social, cultural, and economic forms succeeding one another with what seems to be an increasing, even dizzying, momentum. At any one time, it can be expressed by huge variety in those same forms: different political systems

TABLE 4.1. The Antinomies of Postmodernism

Modernism	Postmodernism
Grand theory	"Local" theory (i.e., theory that is context and group specific)
Truth as universal	Truth as contingent
Truth as inhering in the real	Truth as inhering in discourse
Objectivity	Meaning
Totalizing	Fragmenting
Order	Disorder
A center	Decentered
Homogeneity	Difference
Indigenous	Hybridity
The general, the universal	The specific, the particular
Generalizing	Contextualizing
Essences without history	Concepts with history
Things	Relations
Interests	Identities

and ideas, different consumption patterns, different ways of earning a living, and so on. The aim of modernist thinking was to seek out the universal, the orderly, the general principles underlying the fleeting, the general conditions shedding light on the eventful. "Post" thinking has poured cold water on that endeavor. Rather, the heterogeneous, the fleeting, the eventful, the particular, the disorderly is all one has. All knowledge is specific to particular times and places. Grand theory or what have been called metanarratives, since they tell a story and typically one of progress, are out. Universal truth is a conceit and a costly one at that.

The discourses that mediate knowledge and through which it is expressed are now the object of study. Our knowledge of the world, how we represent it, depends on the application of an inherited body of concepts, categories, and metaphors, each drenched in associations. They are analyzed in terms of their unvoiced assumptions, what they highlight and what they are silent on. Every social theory is partial, situated in particular geohistorical circumstances, and reflecting power relations: in short, a discourse. As a discourse, social theory subjectifies, and through its particular emphases and silences, it grounds practices, which create their own truth. Social theories are interpreted as representations of partial truths that are gendered, raced, classed, or defining a specifically Western viewpoint. Harvey's (1973, Chap. 4) critique of spatial–quantitative geography as "status quo theory" was very much along these lines, but now the critique is turned against historical materialism as itself representing a power position, as a masculine viewpoint as well as a Western one. Marx emphasized production, or what would have been then the masculine sphere as opposed to the more feminine sphere of the reproduction of labor power in the household. He is also held to have rubbished the non-Western societies of the East, seeing them as stagnant and impossible of development until the arrival of capitalist relations of production brought by the Western colonizers. "Post" thinking, rather, insists on analyses of the social construction of knowledge by particular social groups located in particular time–space contexts. Knowledge is always context specific and reflects what is called positionality.

Accordingly, ideas of universal truth are now suspect. Their claims to universality, whether those of neoclassical economics, patriarchal social science, Marxism, or spatial–quantitative geography, are now seen as at best mistaken, and often ploys—strategies to subordinate people, to persuade them that the way *they* interpret the world is how it has to be, how it is naturally so, and, therefore, that resistance is futile. According to this view, knowledge is constructed, intentionally or not, so that it reproduces identities—as male and female, black and white, for example—and, therefore, a particular set of social relations of inequality underpinned by an institutional order reflecting those discursive categories. Marginalization is often thoroughly intended. Knowledge according to this view is always discourse. Language is fundamental. To refer, as Marxists do, to the ideas of others as "ideology" is to attempt to delegitimate. The European geographic imaginary of a world divided between civilization and of barbarism worked to the same effect.

The discourse of so-called Orientalism is especially pertinent. This was a set of ideas formulated in the 19th century in order to make sense of Oriental societies. A dominant motif in this literature was the counterposing of a Europe that was rational, mature, and normal with an irrational, backward, and depraved Orient. Such a discourse clearly flattered Europeans and reminds us that we define ourselves through our representations of others. Furthermore, and accordingly, the new approaches emphasizing discourse were born of struggles against hegemonic knowledges, the feminist struggle against patriarchy, the struggles for racial equality against racism, and for the overthrow of colonial and neocolonial relations.

But not only are theories with universalistic claims—so called metanarratives—defined as the source of oppression; they are also seen as impossible, as having ambitions—to transcend the here and now, the transient by means of a more universalistic representation of the social world—that can't possibly be realized. Rather, all concepts, explanations, standards of truth are impregnated, intentionally or not, with some sociohistoric bias. Social scientists that believe otherwise should confront the paradoxical nature of their practice. On the one hand, they assert that humans are constituted by their particular sociohistoric circumstances and, on the other, that they—the social scientists—are not like other people since they can escape their embeddedness by creating nonlocal, noncontextually valid concepts and standards.

A particular aspect of this approach that we need to be alert to is its nonfoundationalism. The claim is that there are no assumptions about the world that we can call on to give our knowledge of it a firm foundation. Rather, human nature, with all the powers and other causal properties attributed to it by historical materialism and other modernist theories, is discursively constructed. People are, as indicated previously, *subjectified*. There is, therefore, no universal truth that can be used to arbitrate between one theory and another. All theories, in this view, are power plays, and all that we can do is to interpret them in terms of broader discursive formations—formations that entrench existing patterns of inequality or that subordinate formations that challenge those patterns.[8]

What, therefore, might we make of these sorts of argument? Clearly, there are important continuities with humanistic geography. The emphasis on meaning and belief, but now given a more social resonance through the idea of discourse, is there. There is also the significance of the contextual and the way it challenges more abstract, generalizing forms of knowledge. Humanistic geography had been particularly taken up with the technocratic impulse of so much of the theory of the 1950s and 1960s, including that of the spatial–quantitative revolution, but the "posts" have worked to broaden that particular critique as well so as to give it a much sharper political edge. Power is acknowledged to a degree it had not been in humanistic geography, at least in its more programmatic statements. But there are also major discontinuities. Postmodernism and poststructuralism are profoundly antihumanistic. The human as an irreducible, intentional subject, endowed with powers of conceiving the world and of implementing those conceptions practically is considered an impossibility. The subject, rather, is the prisoner of discursive structures that are highly particularized, the creations of particular wills to power, and that can have no universal relevance.

Second, it has introduced a whole new series of buzz words and ideas into human geography and, for some at least, a fertile new slant. Attacks on essentializing approaches at one pole have been complemented by references to the importance of social construction at the other. The idea of the spatial imaginary has entered in the wake of the interest in postcolonialism. It has provided important theoretical ballast for feminist geography. It has also, however, come under some critical scrutiny. There is for a start a certain incoherence about what is being proposed. Grand theory is attacked, but one could equally claim that the "posts" represent similar attempts to find an order in the modern world.[9] Universal truth is abandoned only for the theorists of postmodernity to try to persuade others of their own truth claims. There is a questionable reduction of knowledge to discourse as if how something is expressed determines what is expressed; as if, that is, new metaphors, new words corresponding to new practices, were an impossibility. Historical materialism has been in the line of fire of the "posts" and has not been at a loss in developing a response (Callinicos, 1989; Harvey, 1989b). Among other things, it has wanted to situate the "posts" in their own terms (i.e., sociohistorically) but more specifically in terms of the contradictions and dynamics of capitalist development.

Some of this is obvious. The incorporation of African Americans and women into wage work gave impetus to what became known as New Social Movements and so gave support to the formative significance of "difference" and the construction of identity. A decolonization that was de jure but not de facto was a condition for the discovery of further differences and inquiry into their construction. Some of it is less obvious. As Harvey has argued, the categories of both modernist and postmodern arguments are "static reifications imposed upon the fluid interpenetrations of dynamic oppositions" (1989, p. 339). Capital is indeed fragmenting, differentiating, individualizing, and decentralizing. But it is simultaneously unifying, homogenizing, socializing, and centralizing. At the same time as capital socializes, unifies, homogenizes, and centralizes at the global level through new rounds of globalization so it is eroding the ties of solidarity among workers creating new senses of individualism. The homogenization of experience through the breakdown of the barriers of space allows new geographic differences to be exploited by capital. Capital centralizes control of the global economy in the form of the multinational corporation at the same time as it pressures governments to decentralize wage bargaining structures. So, on the one hand, the forces of what Harvey has called time–space compression condition the cognitive experiences out of which new understandings can be wrought: the sense of geographic difference, of a labor movement now fragmented, among other things. On the other, through the sense of insecurity that those forces engender, the sense that nothing is standing still, people try to make sense out of it all in a way that will protect them from the uncertainty and existential challenge. In that regard, the "posts" are an ideology for the times.

On the other hand, there is no doubt that Marxist geography or what has come to be called historical geographical materialism has been changed by the encounter. The idea of identity, of the desire for recognition and of the stories told in support of identity, is surely one that can be accommodated and which is, in fact, necessary if one is to understand social movements that are clearly significant from the standpoint of the contradictions engendered by the accumulation process: not least the fact of nationalism and the divisions among the working class, which make presenting a united front that much more difficult. That does not mean that one has to accept the interpretations of those fragmentations offered up. Nationalism and feminism have a material base, but to grasp the sheer intensity of national emotions in particular, it is a case of being necessary but not sufficient. The search for recognition has to be harnessed to the cause. And by the same token, of course, that means that historical geographical materialism cannot ignore the idea of discourse, nor the way in which it is used in the course of resolving capital's contradictions.

Historical geographical materialism apart, we don't need to embrace the "posts" in their entirety in order to see that they have some point and can exercise a useful, even important, sensitizing role. We need to be more aware of how our theories might well be serving some political purpose and how little we know of how they might—or might not—apply to the concrete circumstances of developing countries. There is also something to be said for a search not so much for order in the way we have been encouraged but for *dis*order. Geography focuses on spatial juxtapositions, but there are many in the world that we pass over because they seem not to fall into any sort of pattern and, hence, as in the human concept of cause, are identified as causally of no interest. On the other hand, it is precisely these odd juxtapositions that may make a difference to what happens and hence in our understanding of change by, for example, creating the possibilities for technological or social novelty.

In Practice

Whether one might wave the "post" flag self-consciously or not, there can be no doubt that what might be called "post" thinking has had serious and important effects on the work of human geographers. Ideas of social construction and deconstruction, discourse, normalization, and identity have become commonplace in the writing of human geographers. The ideas are often used in ways that might be regarded as eclectic and even uncritical. But even then, and often at the same time, they have been drawn on in very creative ways, producing new viewpoints from which to consider particular human geographies. In this section, I examine three instances of this.

Insiders and Outsiders and Spatial Identities

A fascinating article by Susan Smith (1993) exemplifies "post"-influencd creativity. Her immediate focus was on what might be called spatial identities: specifically a conflict between insiders and outsiders; those identifying with a particular locality and those identifying with more cosmopolitan, big city, values. These spatial identities were, in turn, embedded in very different discourses, the conflict over which generated an abortive politics of difference. Smith's study concerned an unfolding series of events in Peebles, a small town in Scotland about 24 miles south of Edinburgh in what is known as the Borders region.

The immediate cause of controversy was an annual event: the Beltane Festival. A major element of the festival is a street parade featuring people in fancy dress: cross-dressing, exchanging status, and, most significantly, wearing golliwog costumes. For many years golliwog dolls were a pervasive feature of British culture; few who were children more than 50 years ago would not have encountered them as cherished possessions. But as caricatures of blacks, with dark skin, frizzy hair, and prominent red lips, they later acquired the opprobrium of civil rights groups and, it is fair to say, became seen as symbols of a racist culture. It was only recently, however (early 1990s), that their use in the Beltane attracted critical attention. What triggered it were the complaints of an Edinburgh schoolteacher, formerly resident in Peebles, to schoolteachers in that town, who then made a public stand against the practice. This injection of some opposition into the preparations for the Beltane was taken up by the local media and subsequently by the national media, both newspapers and television. In the latter instance, at least the interpretations were critical and did indeed identify the practice as racist.

The dominantly local response and that of the organizers was one of resistance. On the one hand, they defended the use of the costumes in the Beltane as entirely without racial significance. On the other hand, there was an anti-outsider rhetoric, a defense of a local way of life and local traditions. This was concretized not so much by the continued use of the golliwog costumes—indeed there were more of those than usual—but by a systematic drawing of attention to the criticism that had been made of it. In this regard, one character in the parade dressed up as the Edinburgh schoolteacher in question with a fishing net over her (his?)[10] shoulder carrying a captured golliwog.

What seem to have been at stake here are, as I indicated previously, spatial identities, which were in turn entailed by very different understandings of the world. On the one hand, and on the part of those who had lived there for many years, there was an identification with Peebles and an uncritical view of practices that had long been traditional and seen as an essential part of the festivities. The festival was always an insider affair: a celebration of the town of Peebles and its community from which people drew a sense of self-worth. Peebleans celebrated their community feeling, their one-ness, and their difference from non-Peebleans. On the other hand, Peebles had been changing. Like a number of small towns in the Border region, Peebles had been undergoing quite profound social change for some time. Unable to hang on to its young people through lack of jobs, it had on the other hand attracted a considerable number of Edinburgh commuters and retired people. Outsiders, in other words, had come to live in Peebles and, in all likelihood, brought very different values and different understandings of the world with them. The antiracial character of the national media, largely English, had then come in to reinforce the criticism and to underline Peebles's moral isolation. This, however, only intensified the locals' identification with the town. Peebles people were being criticized, made to feel bad about themselves, and unjustly, by an outside world based partly in the cities of England with their racist politics and partly in the cosmopolitan world of Edinburgh with its middle-class values. Their response was to reassert their difference from the English and the lowland Scottish from around Edinburgh. Not only were they different, and proud of it, they were not racist either.

* * *

I think it extremely unlikely that an article like this could have been written before 1980 and perhaps even later. The concepts drawn on would have been quite alien. No one talked of identities. What are also at stake and underpinning these local and more cosmopolitan identities are alternative discourses. The meaning of the golliwogs is at the center of these. The golliwog can clearly mean different things. In one instance, it is part of a time-hallowed practice, the racial significance of which has been subordinated to the meaning of the whole festival, which could certainly not be defined as racist. In the other, it represents old racialized attitudes, which have long been at issue for the big city-based national media in Great Britain. Geography is central to this discussion. It is not just that place identities are in focus. It is also the way in which changing forms of spatial interaction, including the growth of a commuter and retiree population with big city connections and the circulation of national media, are constitutive of those identities in new ways.

Discourse and Political Ecology

Ideas of discourse, difference, and power–knowledge (see following discussion) have found a particularly receptive home in political ecology. The interest in people–nature relations that had dominated human geography in the first half of the century, and then almost dropped out of sight with the spatial–quantitative revolution, revived in the 1970s under auspices that were clearly Marxist in character. Poststructural and postcolonial thinking have provided a strong counterpoint to that early work.

A major theme has been the discursive construction of "nature" and its relation to power. In both lay and academic thinking, nature has long been seen as separate from society. A classic instance is that of "natural" disasters, where particular social arrangements that help bring them about are, in effect, absolved of responsibility. Rather, whether seemingly natural events are "disastrous" depends on the measures taken to offset those effects. Likewise, whether the events are "natural" can be questioned given human interventions into the environment. The critique of the idea of "natural" disasters was already apparent in Watts, as discussed in the last chapter. What has been new has been the emphasis on the discursive constructions at the basis of distinctions between nature and society, the taken-for-granted nature of the latter (i.e., their "normalization"), and their political implications.

I want to discuss an example here that has to do with conservation policy and the discourses underlying it: specifically, the drive, common throughout the developed world, to protect "native" biota, particularly animals. This is, in part, the agenda of conservation biology, sometimes called a pseudoscience, and in part that of governments. Policing "alien" species and restoring "natural habitats" that correspond to some idea of what they were before has been legislated into law in some countries and are in the sights of governments elsewhere.

The vocabulary is evocative to say the least: one of "alien" as opposed to "native" species, "biological pollution" by "invasive" species, the creation of "mongrel ecologies" and "ecological imperialism." But the assumptions on which this discourse is based are highly questionable. Simply separating the supposed "native" from the "alien" is far from straightforward. What is "native" depends on scales of time–space. The longer the period considered and the larger the area, the less likely a species is to be defined as "alien." As Warren has argued, Scots pine is native to Scotland but there are many sites in Scotland where it doesn't grow. So is it alien on such sites? In other words, and as he claims, "rigorous definition of native and alien status is frustrated by sliding scales and blurred boundaries" (2007, p. 432).

This aside, the way in which "alien" species are defined as those intentionally or inadvertently introduced by people into an area where they are not "native" is also troubling. This is because it trades on the notion that people are not part of "nature," which is clearly untrue, besides which, what is one to make of "alien" biota whose propagation into an area was *not* mediated by people in any demonstrable way? This sort of colonization is certainly common enough, and ironically, it is often human disturbance of existing ecologies and the creation of new niches through the destruction of existing "natives" that creates the favorable conditions for it.

The critique of conservation biology is one, therefore, of binaries: the difficulty of separating "alien" from "native" or "natural" introductions from "human" ones. As Warren again concludes, "The social constructivism critique has shown that the borderlines that we draw in order to differentiate between oppositional pairings such as natural/unnatural, pristine/degraded and authentic/fake are of our own making and, to a great extent, arbitrary, ambiguous and imaginary. The present review has shown that the native/alien pairing is equally flawed, contingent on the construction of fixed lines through fluid, hybrid spaces" (p. 449).

Just as clearly, this is a discourse that, regardless of the dubious nature of the binaries that it makes use of, has significant political support. Not least, the herbicide industry has been quick to note its commercial implications. Historically, it found support in nationalist circles, as in Nazi Germany and its campaigns on behalf of "native" plants. "Native" has not just ecological but also cultural meaning. It can signify the protection of a landscape that, mythical or not, has been defined as central to what it means to be English, Scottish or, indeed, German.[11]

Postcolonial Geographies

Of postmodernism, poststructuralism, and postcolonialism, it is perhaps the latter that has had the most impact on human geography. This should not be a surprise since it is the one that has the most obvious and direct geographic implications. Geography or rather a geographic imaginary is central to it. It is an imaginary developed by Europeans, about Europeans, and for Europeans, but in their interaction, through empire and colonization with other peoples elsewhere. Central to it is a discourse defining Europeans on the one hand and Africans and Asians and the Indians of Latin America on the other through a set of oppositions: Europeans as the civilized and the civilizers in the world, the vehicles of reason and all that is morally meritorious, the Christian, the organized, the hardworking, the progressive; and, the Other, or those who were sometimes disparagingly referred to as the non-European, as the uncivilized or barbaric, the disorderly, the irrational, sometimes the pagan, embedded in stagnant, nonprogressive societies. It is an imaginary often described as Orientalism as discussed earlier, but it bears strong affinities to Eurocentricity and the view of Europe as the hearth of reason and the point from which universal truth would spread around the world.

It is a particular expression of what Foucault referred to as power–knowledge: the role of a discourse in subjectifying, imparting to people an identity, and that therefore subjected as well as subjectified, or so it has been argued.[12] In this instance, colonial discourse or Orientalism justified European rule and gave them a confidence in ruling that would otherwise have lacked eluded them. Through imperial practice in the colonies, it inducted indigenous peoples into a self-understanding that would result in an acceptance of their colonial status. Through this particular imaginary, it could be argued that the imperial countries were bringing the benefits of civilization to the colonized and that their colonial status was, therefore, to their own advantage. Alternatively, it could be used to justify the continuation of empire: that the barbaric were incorrigibly so, and the racialized nature of the discourse was, of course, *de rigueur*. This could equally be believed and internalized: White people were always right and white people felt justified in the way they treated "the native."

These arguments have been drawn on in human geography in numerous ways that are productive in an explanatory sense if not always in accord with the primacy that they typically grant to discourse and identity over material interests.[13] They were, for example, claims that were put to very material advantage by colonial governments and European settlers. In a fascinating study, Diana Davis (2007) has shown how narratives of native ignorance could be and were used by the French colonial authorities in Algeria to justify dispossession and the transfer of land to French settlers. Subsequent to the French occupation of Algeria in 1830, French writers constructed a selective history of decline in the area. In Roman times the area had been a major one for wheat production: the so-called granary of Rome. There was also evidence of extensive forest cover, suggesting climatic conditions that would have been entirely consistent with the granary image. But later evidence from the period of Arab occupation starting in the 11th century suggests deforestation and desertification. This, it was argued at the time, was owing to the extensive use of burning to encourage the growth of grass for pasture. Subsequent desertification led to a decline in rainfall and an intensification of aridity. This "history" was then drawn on by the French authorities as an argument for sedentarizing nomadic populations and appropriating land for French settlers who would, it was claimed, be the vectors of "good" land use practice. In these ways, the agricultural productivity of the area would be restored. Yet the "history" was quite wrong since lake sediment core analyses suggest that vegetation and, therefore, climate have actually been stable for the last four millennia.

In colonial India the goal was not so much dispossession as it was maximizing imperial revenue. Indigenous land use practices were regarded as an obstacle to that. Accordingly, one question was how to manage forests for maximum productivity? In classical colonial fashion, the expertise of European forestry was defined as superior to indigenous knowledge (Jewitt, 1995). Any obstacle to commercial yields like shifting cultivation or grazing was opposed; the same applied to the gathering of fuel wood. Particular hostility was reserved for shifting cultivation on the grounds that it was a "lazy" practice and mired the natives in lower levels of civilization.

Algeria, India, and the vast number of formerly colonized peoples are now, in formal terms at least, independent. The discursive power of the West, though, continues. In more immediate terms, this is through the role of the development expert hired by Western development agencies or nongovernmental organization; and in the more enduring terms of colonial, and now postcolonial, identities long in the making. In an article on the Ivory Coast, Bassett and Zueli (2000) refer to what they call a "regional discursive formation" common among development experts working in West Africa. Much like the narrative described by Diana Davis when writing of Algeria, this links supposedly poor land use practices to degradation of the rangeland, then to deforestation, and finally to desertification:

> The image of an increasingly degraded wooded savanna giving way to a grass savanna and, ultimately, desert-like conditions not only persists in the minds of environmental planners, journalists and public officials but is also firmly implanted in the perception of Ivorian environmental and nongovernmental organizations. (2000, pp. 76–77)

The "solution" is the protection of remaining forests from grazing and the introduction of private property rights. In these ways, the development agencies draw discursive power from contemporary arguments about the virtues of sustainability and neoliberal policy. The problem is, though, that the "regional discursive formation" turns out to be wrong and that far from promoting deforestation, burning and grazing pressure actually promote tree growth.

This, however, is too broad a treatment of postcolonial geographies. There were and continue to be subtleties and even qualifications to the argument. To return to Jewitt's work on British forestry policy in India, what she shows is that the superiority of Western knowledge and technique was not necessarily taken for granted by the colonial civil service. Rather, field workers, as one might have expected, were much more likely to see the rationality behind indigenous practice and some did indeed challenge the aversion to shifting cultivation characteristic of more remote colonial authorities. In other cases, Western knowledge was simply rejected. As Carswell (2006) has demonstrated in a discussion of colonial soil conservation schemes in East Africa, the responses of the colonized to policies of terracing, withdrawal of land from cultivation, and livestock culling were far from homogeneous and often hostile, feeding into emergent national movements in the cases of Kenya and Tanganyika. An interesting aspect of her study is the way in which the colonized developed their own interpretive frameworks pinpointing the Europeans as the problem rather than traditional African practice. In Kenya, these understandings embraced land appropriations by Europeans and how it was that, rather than poor agricultural practice, that had led to soil depletion problems. Certainly, as the colonial agricultural officers claimed, there were soil erosion problems, but these were due less to poor practice and more to an inability to settle on land claimed by Europeans.

A different slant on postcolonialism is given by the fact that Orientalism had and continues to have its own more detailed geography. Colonized peoples were regarded as inferior, as lacking the qualities of European civilization but some were defined as more lacking than others. The most obvious expression of this is Africa as uniquely the Dark Continent (Jarosz, 1992). On the basis of notions of different levels of civilization, a hierarchy of the colonized was often invoked in colonial practice: employment in the lower levels of the civil service and even in the distribution of whatever political rights were allowed. In the African case, the presence of Indians and of mixed-race peoples, like the Coloreds of South Africa, has complicated postcolonial politics, with Africans justifying a discriminatory politics in terms of their supposed indigeneity (Mamdani, 2001).

Imperial classifications and claims of indigeneity are also at the heart of ongoing struggle in Malaysia, as Alice Nah (2006) has related. She points out that who is recognized as indigenous can affect who has superior claims to whatever power is going. Indigeneity, however, is socially constructed. In the colony of Malaya, the British chose to work through local sultans as representatives of peoples eventually defined as "Malays," who were then further defined as occupying the "Malayan peninsula." This decision, in turn, was based on the view that the Malays were a more civilized people than the aboriginal inhabitants of the interior, even though the category "Malay" was one imposed on people for whom it would have lacked the clear meaning it had for the British. This sense of unity was stimulated further by the arrival in the colony of Chinese and Indians. In this way, and unintentionally, a majority was created as the politically dominant group in a postcolonial Malaya (and subsequently Malaysia) but one that would draw its legitimacy to form the successor state on the basis of claims to be indigenous to the peninsula. In turn, this belief has formed the basis for affirmative action policies that favor the Malays over other groups, like the Chinese and the Indians.

In the meantime, the aboriginals of the interior have constructed a very different historical narrative, and one for which there is considerable supporting evidence. According to this, if anyone is truly indigenous to the peninsula, it is them, and those who now identify as the Malays were relative latecomers and can be defined as settlers rather than, in their own national myth, the colonized. This has evidently been of serious concern to the Malaysian government, and they have been forced to adopt a discourse of a distinctly Orientalist character. The statement of a former Prime Minister of the country is indicative:

> There could be no doubt that the Malays were the indigenous people of this country because the original inhabitants did not have any form of civilization compared with the Malays. . . . [These] inhabitants also had no direction and lived like primitives in mountains and jungles. (Nah, 2006, p. 291)

One conclusion that might be drawn from this example is that the retreat from colonialism has done nothing to eliminate classificatory practices that worked in their day to the advantage of the colonial authorities.

In this instance, there is a clear local presence in the form of the Malays that benefits from this Orientalist discourse. But Orientalism also survives at the larger scale at which it was originally promulgated; only now, as we saw, the imperial metropoles have been transformed into the West as the source of the modern, of enlightenment, of the normal, of all that is desirable. Postindependence elites have been seduced by the promise of "development," and colonial policies, like the forestry policies that the British introduced into India, endure with all the hardships they thus impose on the masses (Jewitt, 1995).

A BALANCE SHEET

By the early 1990s, human geography had been utterly transformed. The transformation started with the spatial–quantitative revolution and its insistence on the importance of theory and method, something quite alien to what had gone before. This was the basis on which human geography could move forward in equally if not more revolutionary ways. Theory and method were foregrounded but the spatial–quantitative work had its own limits: limits that became increasingly apparent as the technocratic mood of the 1950s and 1960s came under critical scrutiny. As a result, by 1990 a very different human geography was in place. This was one that embraced social theory in far more critical forms than those that had been implicit in the spatial–quantitative revolution. To use the singular, though, is misleading, for it was a question of theories of a very diverse nature: not just the materialism of Marxism, therefore, but also the idealism of the "posts." Regardless, there were commonalities in the reference points for debate: structure and agency, meaning and practice, particularity and universality, place and space, the essentializing and the nonessentializing.

This is to put something of a gloss on the progress that has been accomplished.[14] Very serious tensions emerged and remain. Postmodernism, poststructuralism, and postcolonialism were in numerous respects fundamentally at odds with the more materialist stream of theory in human geography, and particularly with historical geographical materialism. The rejection of materialism and the existence of a reality beyond thought, or at least one that could be corroborated in some way, were fundamental; likewise, the claims to universality represented by historical geographical materialism. With fundamental assumptions so incompatible, one consequence was a high degree of polarization between what emerged as two distinct camps. This is something that has deepened over time.[15] To the extent that there has been some meeting of minds between the earlier social relevance interest and the "posts," it has assumed the form of a bland mix focusing on two sorts of circulation: that of commodities on the one hand and of meanings on the other—something that would go under the heading of critical human geography and that has tended to lose the connection to production that Marxist geography had been at pains to emphasize.

There were and remain possibilities of constructive engagement. One has been through feminist geography for which the "posts" have proved to be a crucial provocation. As we saw in the previous chapter, early work in feminist geography took inspiration from historical materialism and welfare geography. The "posts" were to change that. They were both a promise and a challenge. On the one hand, emphases on particularity and difference resonated. The claims of universality from a very dominantly male intelligentsia and academic geography community struck them as false, as coming from a particular viewpoint that silenced and at best marginalized. Correlatively, there was a sensitivity to a politics of difference in which women could now be seen as the Other of men, as lacking their, evidently superior, qualities.

Yet from a feminist standpoint there was also something problematic about postmodernism and poststructuralism. The rejection of underlying structures and of the emphasis of the grand narratives of progress bequeathed by the enlightenment was a real concern. This is because it seemed to curtail the possibility of female emancipation. Similarly discouraging was an emphasis on subjectification, which seemed to imply a discursive prison from which women could hope for no escape. As Massey concluded:

> One of the main attractions of the postmodern perspective would seem on initial viewing to offer the prospect of a greater democracy through its recognition of the reality of a variety of viewpoints, a plurality of cultures. This has its underside: those viewpoints and cultures may, for example, run counter to what we have been accustomed, from a modernist perspective, to think of as progressive, and postmodernism forbids us from evaluating. (1991, p. 32)

One of the ways out of this dilemma of particularity without the prospect of emancipation has been to argue for a separate structure of social relations: patriarchy, which incorporates conceptions of difference and is subject to redefinition through struggle. It has, therefore, been common to claim that gender relations exist independently of class and cannot be reduced to class relations.[16] The latest incarnation of this is what has been called "intersectionality." This pluralism is not always argued out, so that it has often remained unclear whether one is talking about a structure of material relations, as in a division of labor, or of identity, or both.

The response of historical geographical materialism has likewise not been entirely negative. One of the effects of poststructuralist arguments has been to shed critical light on the Marxist concept of ideology. For Marx, ideology was to be contrasted with science. Science was truth and his approach was the scientific way, while ideology was an understanding of the world based on illusions. Ideology was based on a world of appearances, while science laid bare the real conditions for human practice, including the ideological forms through which people in class societies made sense of those practices. For the "posts," on the other hand, the surface is all one has. The truth of things is a result of discourse. Marx's depth ontology is rejected since there is no essence that can shed light on a world of appearance. This might seem a weak base for rapprochement. The problem is that Marxists have been deeply affected by the antipositivist critiques of the 1970s (see Chapter 6). The prioritizing of sense data will no longer suffice since it is now recognized that we can only have access to the real through concepts. This has shaken confidence in the old science/ideology distinction and discourse has become an acceptable way of threading the needle. But discourse is now incorporated as an aspect of the social process as a whole and not as the condition, if not the determinant, of social life as in poststructuralism; and moreover, of a social process at the center of which is production (Harvey, 1996, Chap. 4).

Beyond these limited possibilities of cross-fertilization, there is a deep sense of impasse. Both sides try to make sense of the other on their own terms. I discussed earlier in the chapter the way in which postmodernism defined historical materialism as one of many grand narratives reflecting a narrow white, Western, and for some feminists, masculine will to power. For Marxist geography there have been two responses. One has been to situate the "posts" in the contradiction-ridden development of capitalism: as an approach to the world that captures and responds to the ever fleeting nature of experience that is part of what Harvey has defined as time–space compression (Harvey, 1989). The other has been to show how Marxism has always recognized the particular and the ideal but as moments of universal tendencies grounded in people's material engagement with the world through production. Accordingly, language can certainly be accepted as crucial to how people make sense of the world but, as Marx and Engels made clear in *The German Ideology*, as an aspect of the labor process. And as such it is always contested, and class has been central to that contestation (McNally, 1995). Likewise, to argue for a continuing development of the productive forces under capital, which seems difficult to deny by even the most dogmatic antiessentialist, is not to ignore the fact that capital's concrete trajectory can never be predicted. The particularities of development, particular products, particular labor processes and skills, particular raw materials, are always up for grabs (Lipietz, 1986). But the divisions continue, not least in how human geography's key concept, space, is to be understood, as we see in the next chapter.

NOTES

1. See *Society and Space*, Volume 5, Number 4, 1987.
2. According to Entrikin (1976), the formalization and advance of humanism in Anglo-American geography sprang from a deep dissatisfaction with the "new geography" of the 1960s and its concerted reformulation of the discipline as spatial science. Humanistic geography shared in the critique of positivism and was for a time represented as "a form of criticism" through which "geographers can be made more self-aware and cognizant of many of the hidden assumptions and implications of their methods and research" (1976, p. 632).
3. There are numerous examples here: Harvey, of course, but also, and in no particular order, Doreen Massey, Ray Hudson, Michael Webber, Eric Sheppard, Richard Peet, John Holmes, Peter Taylor, and left-leaning people like David Smith and Dick Morrill.
4. This was published prior to his book *Outcast Cape Town* (1981) on the same topic.
5. Entrikin's (1978) interpretation of this particular version of humanistic geography is, therefore, quite apropos. See Note 2.
6. Compare Sayer (1989a, p. 309): "The social is not reducible to the individual. Even individual subjectivity is structured through intersubjective relations and expressed in socially available concepts and language. Consciousness never consists of presuppositionless understanding, nor can it be reduced to expressions of private beliefs; for it always has structure or syntax, whether linguistic, conceptual or artistic, and such structures have a certain autonomy."
7. One approach to such a reconstruction was outlined by Sayer (1989a).
8. An excellent discussion that examines nonfoundationalism can be found in the form of a debate between Chomsky and Foucault at *www.csudh.edu/dearhabermas/chomfouc01.htm*
9. Driver (1992, p. 33) recognizes this problem in his discussion of Orientalism: "Postmodernists would thus interpret Orientalism as one of the master narratives of modernity. They are faced, however, with an immediate problem: how is it possible to write a critical history of Orientalism which avoids the very essentialism it seeks to expose? For Said's own account of Orientalism might itself be (mis)conceived as a grand narrative, drawing local and individual differences into its vision of a governing structure of cultural power." Driver is clearly not happy with this response and wants to counter it, which he does by appeals to the heterogeneity of Orientalist writings and the need to ground them in very diverse contexts. I find this disingenuous and an odd refusal to recognize the interdependence that would surely have brought the different literatures together. Compare also Harvey (1992) in responding to criticism of his book *The Condition of Postmodernity*: "Postmodernists . . . cannot criticize *The Condition* as wrong, misguided or fundamentally misconceived without deploying truth terms of their own which presuppose they have an ultimate line on a truth they theoretically claim cannot exist" (p. 322).
10. The teacher in question was indeed female but given the practice of cross-dressing in the parade it is unclear whether it was a man or a woman performing the caricature.

11. A more recent example does indeed come from Scotland, where the replacement of "alien" tree species by "native" ones has been understood as part of a process of national regeneration (Robbins and Fraser, 2003). The Caledonian Forest, which supposedly covered large parts of the Scottish Highlands, retreated under the successive impact of cattle, sheep, and later deer forests. Later still, state-sponsored re-afforestation for the purposes of stimulating a commercial timber industry drew on more productive, faster maturing "alien" species. With the increasing influence of Scottish nationalism, not only is Scottish history being retold, but also what are regarded as historical landscapes that existed before the union with England are being looked at as worthy of reconstruction. This means an ecological nationalism in which supposedly "native" trees, like the Scots pine, the silver birch, and the alder are reintroduced in an attempt to re-create the old Caledonian Forest.

12. I have inserted this qualification because I do not think that the subjectification assumed is always consistent with the conditions that would have had to be in place in the colonies for it to come about. Not least, literacy, a necessary condition for imparting particular "truths," was far from common and, given the absence of mass schooling, not something that colonial governments changed very much.

13. But see Gregory (1998).

14. I am aware that some, particularly those of "post" persuasion, will be uncomfortable with the reference to progress. But as in other respects, they are self-contradictory on this since, in claiming that theirs is a better way of understanding the world, they implicitly embrace a concept of improvement.

15. The journal *Antipode* was from very early on in its history the mouthpiece for Marxist geographers. By the first decade of the new century, this identity was looking distinctly frayed as "post" thinking became more apparent. The result has been the creation of a new journal, *Human Geography*, for those Marxist geographers no longer comfortable with *Antipode's* shifting center of gravity.

16. See Chapter 6 for a discussion of this point.

CHAPTER 5

New Understandings of Space

CONTEXT

As we saw in Chapter 2, the spatial–quantitative revolution entailed a dramatic shift in the way that space was understood. The earlier space of regional geography and people–nature relations had been one in which relations over space, what the spatial–quantitative geographers would come to describe as "spatial interaction" had been of quite limited significance. It was as if events, land uses, towns, were slotted into a preexisting set of geographic pigeon holes between which there were only limited relations at best. Space seemed to exist in and of itself, in the form of coordinate intersections on a map, regions, countries, or the continents. It seemed to be independent of anything occupying it—so many containers waiting for things to happen, to be discovered.

The position of the spatial–quantitative geographers, whether they thought about it or not—and for the most part they didn't—was quite different. Their focus was on movements and connections over space and, the ultimate objective, how those movements and connections helped account for locations like those of cities or agricultural zones. The walls of the pigeon holes of absolute space were now transgressed. Events were no longer assigned to coordinate intersections or places, each of which was unique; rather, they were assigned to points in space defined relative to one another—points that were more or less distant, more or less in the same direction, more or less accessible—and so substitutable one for another in a way that was impossible with spaces as absolute in themselves. Space now existed only because matter existed in particular geographical arrangements: To talk about the friction of distance was to talk about the costs incurred as a result of friction with the underlying surface.

This is a highly simplified history. Relative space did not wait for the spatial–quantitative revolution to make its entry. Notions of urban spheres of influence were already evident in human geography prior to the 1960s. The concept of diffusion is central to some of Sauer's work as in his utterly seminal 1952 *Agricultural Origins and Dispersals*. Even so, the spatial–quantitative revolution made a major difference to how the majority of geographers thought about space. The race was then on to uncover the multiple ways in which what was called relative location made a difference.

With the increasing awareness that a human geography with limited attention to social theory, particularly of a critical sort, was intellectually indefensible, this would have to change once again. Conceiving space as relative could often have deterministic consequences: space or more accurately spatial arrangement could be determinant of where people shopped, worked, what they grew on their farms, whether or not they had jobs. As Massey noted (1984, Chap. 2), one geographical distribution was explained in terms of another. This was the guiding conceit of the urban and regional planning for which many human geographers were being trained.[1]

Putting social relations at the center would necessarily entail a different conception of space. They were now seen to condition space relations; capitalism resulted in certain sorts of spatial arrangement as did patriarchy. This, however, risked replacing a spatial determinism with one of a social kind. The upshot would be to, in effect, recognize that the social necessarily incorporated a spatial moment, not just in the doing but in what it was from the start: that social relations could not possibly be anything other than spatial relations. One not only produced space in and through sociospatial relations, one was sociospatial in one's very constitution. In Harvey's (1973, p. 13) words, it was "space regarded . . . as being contained *in* objects in the sense that an object can be said to exist only insofar as it contains and represents within itself relationships to other objects."

Once again, there were precedents. For a start, the classical political geography of Mackinder and Bowman, with its emphasis on policy, its alertness to the changing geographical circumstances of Great Britain and the United States, respectively, and what that implied for their very constitution, could not be anything but. Hägerstrand's Monte Carlo simulation studies had similar qualities to them. A reason for this was the way in which they all had a strong sense of the importance of time in understanding human geography; how events took place not so much in space, and relating to the existing arrangement of things but in time–space, and this, of course, required one to think about human geography as something always in the course of becoming.

In brief, with the advent of social theory of a more critical sort, social and spatial relations were to be brought together and result in the domination of concepts of relational space. But quite *how* they would be brought together would depend on understandings of the social. The more materialist conceptions of the social process instantiated in historical geographical materialism lent themselves to quite different interpretations than the "posts" of the late 1980s. The production of space here assumed a sharply material sense as in the production of Massey's spatial divisions of labor and subsequent geographically uneven development. Yet the sort of production envisaged, in say, Susan Smith's discussion of Peebles was of a clearly discursive sort and one in which the media were active agents.

UNDERSTANDING RELATIONAL SPACE

In the introduction to his book *Social Justice and the City*, published in 1973, Harvey briefly identified and explained the different meanings that space might have: space as absolute, space as relative, and space as relational. As far as I am aware, this was the first time that these distinctions had been made in the human geography literature. Clearly absolute space had been implicit in the old regional geography of the earlier part of the century, and the spatial–quantitative revolution had drawn on a concept of space as relative. But now they were made explicit. Furthermore, relational space had made its entrance, and this was where the future of human geography would lie, even while continuing to draw on the other understandings. Nevertheless, making sense of the implications of relational space, and therefore exploiting its potential, was not immediate. It was now recognized that space had to be seen in relation to social relations; that spatial relations *were* social relations. This, though, was an abstract understanding and would take some time to work out in terms of its concrete implications.

The initial tendency was to draw attention to how social relations entailed certain sorts of geography without an investigation of how space made a difference to the social process. Clearly, social relations conditioned space relations. Marxist geographers explored the implications of capitalism in this way. The tendency of capitalism was to produce an industrial reserve army, and this was apparent geographically in what Dick Walker (1978b) called "a lumpen geography of places." Feminists identified gendered geographies and in particular the differentiated relations of women and men to living place and workplace, respectively. What remained unclear was how space was to be incorporated as part of the social process.[2] Clearly, human geographies reflected dominant social relations; but how did space relations make a difference to those social relations? In what sense could the social process be conceived as a socio*spatial* process? Up until then it seemed that people in their social relations were the active subject while space was the object. So how could space be causally significant? By the early 1980s it was becoming clearer. The 1985 collection edited by Gregory and Urry was particularly important in this development.

One of the answers in that book was that space was for the most part of only contingent significance. This was the response of Andrew Sayer (1985). While he recognized that what he called in the language of critical realism, causal properties, had significant spatial aspects—people needed space for their activities, they had the power of mobility and so on—this did not amount to very much in the understanding of concrete geographies. Necessarily human geographers would have to be consumers of theory from elsewhere in the social sciences. The other answer was more reassuring and occurred initially as a spatialization of structuration theory. Among all the contributors to that book, this is most explicit in Soja's chapter, though implicit elsewhere, particularly in those of Massey and Harvey. As Soja claimed, "As a social product, spatiality is simultaneously the medium and outcome, presupposition and embodiment, of social action and relationship" (1985, p. 98). Space was a necessary aspect of social relations, both a necessary and reproduced/transformed condition of social action and to be theorized as such.

This, of course, was very far from the formulations of the spatial–quantitative geography of the 1960s, and allowed a recuperation of some of those aspects of human geography that it had, in effect, served to silence and delegitimate. A major case in point was the region or, more generally, the idea of place. Spatial–quantitative geography had privileged space since this was the realm of the generalizable: of substitutable locations, of trade-offs between different location factors, of location pattern or spatial organization. Regions were unique and hence resistant to its law-seeking, theory-building strategies. This, it turned out though, was only the case if one adhered to space as relative. Relational space implied something else. As Doreen Massey drily suggested in 1983, "One can do more with the unique than contemplate it" (p. 75). Places entered into the sociospatial process through their particular, distinct qualities and the functions that those qualities might perform in that process. Places, like human geographies in general, were socially produced. In studying them in this way, one could explore the ways in which the production of space actually worked out in particular cases; how places got transformed so as to alter the contours of the spatial division of labor and how particular qualities of places were mobilized to function as crucial aspects of the circulation of capital. Places were made but, through their unique qualities, their emergent properties, they also "made" in their turn; they were sites of innovation. As Storper (1987) averred, "The local makes the global."[3] Hitherto there had been talk about a "new regional geography" or at the least, the need for one. Now it seemed to be a real possibility.

Materialist Understandings

A useful entry point in talking about the production of space is through the concept of the socialization of production and reproduction. Whether one subscribes to a Marxist view of the matter or not, it is plain that modern capitalist development depends on a massive socialization of how things get produced and equally how the people who do the producing, or will in future years, get reproduced. The evidence is very clear. In trying to understand the revolution of production in his time, Adam Smith emphasized the division of labor and to a considerable degree he was right.[4] Production increased through specialization; through the division of production according to what was being produced, but also through the division of labor within the factory or mine. The division of labor worked in part through the sharing of particular means of production, and this is a second aspect of the socialization of production; not just the collective working of massive means of production in the factory like an assembly line but also the sharing among firms of a transport network, airports, and docks and the physical and social infrastructure of cities. Similar comments can be made about social reproduction. But this socialization of production is just as clearly a socio*spatialization* of production. Divisions of labor are *spatial* divisions of labor. Collective means of production and reproduction like cities have geographies, which, in turn, condition spatial divisions of production and reproduction alike. This has been very clear in the work of human geographers on the problem of geographically uneven development and how it comes about.

The name that comes most readily to mind when thinking about work on spatial divisions of labor is that of Doreen Massey. Her book of the same title (1984) had a remarkable impact on the field. There had been earlier premonitions of the arguments she developed there. In an important paper in 1979 and in the course of discussing the changing nature of the regional problem, she distinguished between older and newer spatial divisions of labor. Older spatial divisions of labor referred to geographic specializations in what was produced: an older geography of coalfields, manufacturing belts, cotton belts, and the like, a geography instilled into generations of British schoolchildren as they learned about cotton textiles in Lancashire, metal goods in the Black Country, and the woolen district of West Yorkshire. What Massey remarked, though, was something new. Less a specialization in terms of product, this was more one based on position in the technical division of labor. It might be within the structure of a single firm; or connect one firm to another, as in relations of subcontracting; or both. According to this, firms might have their production in one location, their research and development in another, and their headquarters yet elsewhere. Or again, production might be subdivided into less skilled and more skilled, again located in different places.

In other words, the labor process had been broken down into different stages, each characterized by different labor demands, and each stage allocated to different places in such a way as to maximize profits. In the British case, firm headquarters might remain in London in virtue of accessibility to air transport, to banks specializing in their line of business, and to marketing firms or simply because of advantages in keeping the corporate ear to the political ground at Westminster. Research and development might be allocated to some smaller town in southeast England, not just because of class structures that meant access to appropriate workers but also because the area was attractive as a place in which to live and so made it easier to recruit workers who were very much in demand. Semiskilled assembly work, on the other hand, might be consigned to a declining coalfield area where there were large reserves of female labor willing to work for a substantially lower wage than elsewhere in the country. Meanwhile, the more skilled work might remain where it had always been: where the firm had originally grown up and built up a skilled workforce, difficult to replace when relocating elsewhere.

The implications for the production of human geographies were evident. Instead of a landscape and language of coalfields, textile and chemical belts there were now headquarters cities, branch plant towns, along with those retaining a monopoly of the skills-intensive parts of the labor process: jet engines in Hartford, Connecticut, and Derby, England, and machine tools where they have always been—in towns scattered across northern Illinois and southern Wisconsin, from Rockford in the southwest to Milwaukee in the northeast. The story behind this was indeed one of production, and production in several senses. Typically in the changing historical geography of an industry, there had been changes in the labor process that allowed some geographical dispersion of different parts of the technical division of labor; and then the geography of employment had been transformed. With the deskilling of the rubber tire industry in Akron, Ohio, it had been possible to relocate the actual production to small towns in Oklahoma, Mississippi, and Alabama. New industrial functions had been acquired, allowing some southern towns, at least, to grow; while Akron had been given a new role as a poster child for deindustrialization.

In other words, and as Dick Walker (1978b) had emphasized, as the spatial division of labor changed, so capital shifted between regions, leaving behind it pockets of unemployment and producing what he called a lumpen geography of places. The fact of unemployment, or even the threat of it, then changed the class relation by making workers in particular places more vulnerable to the demands of employers: for changes in labor contracts and work conditions in particular. This focus on the capital mobility through which changes in spatial divisions of labor were mediated and its implications for the employment relation was taken up by others, notably Gordon Clark (1981).

Massey's work opened an important window on the geography of uneven development, but she had little to say about cities, except as places that might retain or lose employment as firms created new spatial divisions of labor. Rather, it was Allen Scott who led the way in underlining through a series of papers the significance of the city as a collective means of production, even while he did not use that term. Massey's focus had been for the most part on larger firms and the creation of new spatial divisions of labor largely within the corporate confines of a single firm. One of the changes in economic geography that she emphasized in consequence was the transfer of deskilled labor processes from larger cities to smaller towns, often in areas of relatively high unemployment. There was some emphasis in her work, therefore, on the decline of cities as places of industrial employment (Massey and Meegan, 1978). Scott, on the other hand, while recognizing the fact of new spatial divisions of labor and their locational consequences also identified countertendencies. He underlined the continuing significance of cities as industrial locations largely as a result of their attractions for vertically disintegrated, smaller firms. Clustering together allowed the externalization of production processes to others. This facilitated specialization while also allowing the thresholds to be achieved for new auxiliary functions like the provision of specialized transport, legal, and accounting services. The resultant cost reductions encouraged concentration of particular industries in particular cities and made infrastructural expansions more viable for city governments. Subsequent competitive advantages forced firms elsewhere out of business and further enhanced locational concentration.

To some extent, Scott connected processes of vertical disintegration to dilemmas of firm governance. He saw firms, somewhat controversially,[5] as systems of economic transactions. Optimal firm size was then determined by the relative efficiency of internalizing transactions within the firm as opposed to externalizing them to other firms. But he was also keen to give his arguments a stronger base in the mutations of capital and in particular those identified by regulation theory. The larger corporations at the center of Massey's arguments were typical of fordism. With its decline and the emergence of a new regulatory regime of postfordism, there was a recrudescence of smaller, vertically disintegrated firms catering to what Scott saw as smaller, more uncertain markets in which product design was increasingly important. Firms needed to be more flexible and able to change process and product configurations quickly. To internalize the production of particular inputs was to court disaster.

Rather, they should be bought in as circumstances required. Heightened competitiveness also put a premium on more flexible labor relations. This sort of flexibility was for Scott "typically situated in networks of extremely malleable external linkages and labor market relations" (1988, p. 174). This in turn was encouraging a reagglomeration of industry, not in older industrial centers with their traditions of militant labor, but in what he termed "new industrial spaces": places that were relatively new to industrial work like the much celebrated Third Italy, Silicon Valley, and Cambridge in England.

The interventions of Massey and Scott were immensely important in developing an understanding of the production of space and in many ways were complementary to one another: Massey's emphasis on larger firms was complemented by Scott's on the smaller, vertically disintegrated firm; Massey was excellent on processes of industrial dispersion while Scott deepened the literature on agglomeration processes. Yet both were, necessarily, partial endeavors; glimpses through particular windows without an underlying framework which could make sense of both of them as aspects of capitalist geography. Apart from the occasional reference to labor relations as a condition of location, Scott's was particularly lacking in attention to questions of power. Massey did a little better, trying to position the locational changes set in motion in Great Britain by the adoption of new spatial divisions of labor with respect to national policy, but her interpretation lacked a strong sense of the territorial, which perhaps reflected the highly centralized form of the British state and the way it has tended to counter strong localized impulses of a territorialized nature. In these regards, Harvey's "geopolitics of capitalism," introduced in two groundbreaking papers in 1985, was of huge importance.

In an analysis that appeared in the Gregory–Urry (1985) collection, Harvey laid out a compelling vision of capital's ever changing, always dynamic, historical geography, of which, of course, the changes identified by Massey and Scott were examples. As per the arguments laid out in the discussion in Chapter 3 of his earlier paper on the geography of accumulation, Harvey looked for the source of this dynamism in capital's contradictory character. Capital had to expand, and it expanded in ways that inevitably served ultimately to impede that expansion. There was a constant tendency to what he called overaccumulation: the piling up of surpluses of capital for which there were no profitable outlets. One of the ways in which this contradiction was suspended was through the creation of new spatial fixes: new locational arrangements, the creation and further exploitation of which could soak up surpluses of money capital in profitable ways. Massey's new spatial divisions of labor provided one example, along with Scott's agglomerative tendencies, but so too did the emergence and expansion overseas of multinational corporations.

Now, though, capital's contradictions, seemingly suspended, became manifest in new ways. Capital circulates through different forms: money, raw materials, machines, physical and social infrastructures, and finally products, the sale of which allows the circulation to start once again. In some of these forms capital's circulation is inevitably slowed down. Money capital is embodied in fixed facilities, public and private, of long life; only slowly are those values retrieved through the use of those facilities in production and subsequent sales. Similar arguments apply to labor skills built up over many years and whose value is only returned to the employer over a long period of time. From this viewpoint, cities and countries are fixed resources, publicly and privately owned, which stand to be deval-ued—factories, housing, workers, schools, utilities—unless value can con-tinue to circulate through them. Yet clearly there is no guarantee of this as capital's landscape undergoes constant change under the impulse of creating a new spatial fix through which the problems of overaccumula-tion can be suspended.

In short, the mobility of capital enters into contradiction with the fix-ity that it necessarily assumes at a certain phase of its circulation. But this is a contradiction that is experienced to varying degrees over space: more in Akron as a result of the decentralization of rubber tire production, for example, than in those small towns in the American South to which that production was directed. This is the basis for what Harvey describes as the geopolitics of capital: a struggle between cities, regions, and coun-tries to ensure that the devaluation that is the inevitable consequence of the constant search for a new spatial fix will be visited elsewhere. This is the politics of territorial competition for new investments led by local and national governments; the politics of product development through state subsidy; the constant downward pressure on wages and conditions of work to enhance cost competitiveness and the like; and all discursively packaged and sold.

In short, Harvey was able to link the different moments of the uneven development process together—the political (hence struggle) and the cultural in addition to the economic—while demonstrating the crucial importance of contradiction in understanding the production of space. His paper is of absolutely seminal quality in a case where the term 'semi-nal' is no exaggeration. This is because of the novel ways of understand-ing geographies that it opened up. He explored some of its implications himself when considering the case of urban politics in a paper that is an excellent complement to the work of Allen Scott on urbanization (1985, Chap. 5). The conception also laid the ground for the author in his work on both the politics of local economic development (Cox and Mair, 1988) and the politics of scale (Cox, 1998).

Much remains to be explored, not least in the area of political ecol-ogy. Swyngedouw's (2007) work on Franco's hydraulic revolution in Spain is a striking example of the possibilities. Through a massive program of dam construction and interregional water transfer, the Franco government positioned Spain for a future role in the international division of labor through expanded irrigation and enhanced electrification in a country lacking significant coal resources. The program also had sharply centralizing implications: crucial to countering the country's fissiparous tendencies and girding its loins for struggles around uneven development at a global scale. Producing space in Spain, therefore, was an ambitious project not just in irrigating lands and so promoting agriculture and in providing power for industrial areas, but also and emphatically in the production of geographic scale, a theme to which I return later.[6]

Discursive Construction

Accounts of the construction of space drawing inspiration from the various posts have been radically different. Here, and by definition, there are no grand narratives like capital to structure the understanding; no structures like the division of labor to lend coherence to explanatory accounts; no integration of places over space through some structured agency, whether that of capital or the state; in short, no set of underlying material limits and possibilities to act as check on, and facilitator of, what is possible. The emphasis has lain elsewhere. It is, rather, the irreducible particularity of things and their discursive formation that has attracted attention: their continual formation, dissolution, and re-formation. Things are formed through the communication of meanings held to be true, and hence in fields of power relations, since all claims to truth are claims to power. The universals through which people have tried to grasp the geography of the world are simply so many different instances of power–knowledge. Discourse emerges in the context of particularity, in a context of fragmentation, therefore, that is inevitably spatial. It is different networks of discursive relations that make people in places different from those in other places rather than variations around some national or global norm. The local in the relative sense is crucial, which explains the emphasis often placed on "context" and on "local pockets of order."

Opposed to the particularity of things are the abstractions of the various narratives that seek to impose some unified, totalizing understanding on the world; their divisions and classifications of things in accordance with some of their one-sided aspects and the neglect of others and the creation of some overarching vision of the world on their basis. A common critical focus has been the various binaries that have characterized geographic thought, not to say social thought more generally, and both lay and academic: attempts to construct people in particular places and to create an imaginary geography that affects how people understand themselves, their susceptibilities to persuasion, seduction, and cooptation. In this way, real geographies approach imaginary ones; truth claims are claims to power and claims to power require claims to truth through which to justify themselves.

The binaries are numerous. Cities are divided into "suburbs" and "inner city." "Urban" is opposed to "rural," "metropolitan" to "provincial," "city" to "countryside" and "sunbelts" to "rustbelts," "London and the southeast" to "the rest of the country" and "first world" to "third world." The study of the controversy surrounding the Beltane in the Scottish border town of Peebles, recounted by Susan Smith and discussed in Chapter 4, can be regarded as an investigation of an attempt to impose a particular view of race relations—one emanating from the national media and strongly resisted by those who would surely have been classified by the media, explicitly or otherwise, as "provincial." The genderized understanding of public–private space, the imputation of the home to women and of the public spaces of employment and the state to men is equally pertinent, if different: different because it seems to be a discourse rooted not in places but in positionalities that are more territorialized. White geographies of who is "in" and "out of" place are of the same genre.

One of the most influential binaries in human geography has been that of "the West" versus "the rest." The critique of what is called "Eurocentricity" is well known: the idea of Europe or, now, "the West" as a norm to which the less developed world should aspire. This "norm" includes political and economic institutions, ways of understanding the world embodied in physical and social science, and modes of artistic expression. Edward Said's "Orientalism" was a concerted attempt to reinterpret this in terms that, much as he disagreed with Foucault, approximated his power–knowledge. Through its stress on imaginary imperial geographies, it has become a center piece of the approach to understanding known as "postcolonialism," as discussed in the last chapter.

Orientalism sedimented a particular identity in Europeans: one of superiority and as conveyors of enlightenment. Jansson (2003) has shown how it is a construction that finds echoes in nation building and in creating particular national identities. The example he has in mind is the relationship of the United States to the American South. Through a study of a highly influential book, W. J. Cash's *The Mind of the South* (1954), he has drawn attention to a discourse about the South that works, through claims about difference, to facilitate the construction of a particular and flattering American identity as "what it is" as opposed to "what it is not": "peaceful" rather than "violent," "enlightened" as opposed to "racist," and "energetic" in opposition to the "laziness" of Southerners. The South, in other words, is defined as Other, as in its traits not truly American.

Recently, Simon Springer (2011) has given the imaginary geographies of Orientalism a useful updating. He has drawn attention to the view that, in contrast to the West, some non-Western cultures—African, Arab, Asian—are peculiarly susceptible to violence. Apart from critiquing this view, a critique to which I refer later, he also shows how it has been mobilized by neocolonialist ventures, particularly those sailing under the flag of neoliberalism. Thus, neoliberalism has defined itself as pacifying, by allowing people to realize interests that would otherwise be frustrated and end in violence. It is rational and democratic and democracies are nonviolent, and so on.

Springer's arguments recall earlier ones advanced by Pierre Clastres (1977). Clastres was an anthropologist and not a geographer, but his work is an interesting complement to that of Springer and in some ways adds subtlety, not to say provocation. The binary at the center of Clastres's interest is that relating a variant on the civilized–barbaric one—civilized–primitive—to the presence or otherwise of a state. The standard view has been that civilized societies have states; primitive ones do not. That is the reason that primitive societies are violent. But according to Clastres, the reason they lack states is that they don't want them. States force people to do things they otherwise would not do. They are a means of extracting a surplus from people through taxation and corvée labor, of dispossessing them and forcing them into relations of commodity exchange. The absence of markets does not signify lack of civilization. Rather, so-called primitive peoples often exhibit great technical ingenuity in obtaining subsistence from extremely unpromising conditions. "Civilization" in the form of the state is something that people don't want, which is, according to James Scott's (2009) recent book *The Art of Not Being Governed*, the reason that in Southeast Asia they have fled to areas that the state has historically had difficulty controlling. But in part *contra* Springer, and according to Clastres, primitive peoples at least *are* violent both with respect to each other and any chief who claims to be a chief. This though is to prevent the emergence of a state, to keep power dispersed.

Given these critiques of place-making discourse, particularly that of Clastres, who gives the differences identified in Orientalist understandings a very different and more critical interpretation—yes, he says, some peoples are more violent but that is because they resist the state and, furthermore, they have good reasons for resisting it—how should we understand the construction of space? How are people and their space relations to be understood in all their particularity? One idea that has yet to be mentioned in this discussion is that of hybridity, and this is central to a construction, or more accurately, *re*-construction of space relations. This is of crucial importance. Things are irreducibly hybrid in their formation. That is what makes them particular. They are formed at the intersection of innumerable discursive influences. It is for this reason that people have multiple identities. Totalizing narratives like that of Orientalism or expertise make their own abstractions for their own purposes from that fundamental hybridity. Through emphasizing one side of a binary over another, they assert a particular presence in the world, subordinating and possibly silencing; certainly separating and expounding a knowledge that cannot possibly be true to the world's hybrid nature. If there is anything essential in the world, and the posts claim that there isn't, then it is particularity, and particularity finds its origins in the intersection of multiple discourses.

In the geographic literature, this has been registered in several related ways. One of the first to take this question up was Chris Philo, who, as early as 1994, was calling for what he called "a geographical history." By this he meant making past geographies part of the explanation for major historical changes like the emergence of capital and the formation of racial identities: to understand history in terms of comings together and separations in space but also in time. In an earlier and related paper (1992), he examined what he called Foucault's geography, which, he argued, amounted to a

> geographical way of looking at the world in which one sees *only* "spaces of dispersion": spaces where things proliferate in a jumbled-up manner on the same "level" as one another—on the one level where advanced capitalism and the toy rabbit beating a drum no longer exist in any hierarchical relation of the one being considered more important or fundamental than the other—and on which it can never be decided if "the essential" has been sighted (because there simply is *no* "essential" to be sighted or because, even if there is one, we can never know whether it has revealed itself. (p. 139)

However, pride of place in this discussion of how to understand the construction of space should go to Doreen Massey. Although she would almost certainly reject the label "post,"[7] Massey's more recent work shows how some of the claims made by poststructuralism and postcolonialism can be turned to creative use in human geography, and quite remarkably at that.[8] After Philo she was the first to recognize the significance of space in a critique of grand narratives.[9] Soja (1989) had earlier intimated the connection between what he called the spatial turn in the social sciences away from a privileging of the historical dimension for a geography that was postmodern, but had failed to recognize its true significance.[10] Massey's key insight (1999a) was to argue for the way in which the introduction of space into theories that emphasized the temporal dimension had markedly unsettling effects.[11] Rather grand narratives did not necessarily exclude space. It was just that in their reference to developed and less developed countries, modern and traditional places, or more globalized and less globalized, they diminished its significance by assigning places to slots in a seemingly preordained march into the future: The less developed would follow in the path of the more developed, places characterized by more traditional forms of social practice would eventually be modernized, and so on.[12] The future of places was, therefore, closed.

Particular place trajectories, their arcs of change over time, were normalized. It was "normal" for countries to develop, become modern, secular, democratic, and urban. In this way, grand narratives, by closing down other possibilities in favor of particular forms of social and economic change, had a political purpose: development, revolution, or whatever was in a country's future and that was the only possibility. On the contrary, Massey argued, space was the sphere of difference, and this meant openness to the future. The future was not preordained and closed in the way conceived by grand narratives. Influences from elsewhere could come together in particular places, providing the conditions for novel forms of social structure and practice and change.

In this way, Massey (1992) has also shown how the order identified by modernist social science, the orderliness of the geographies of the spatial–quantitative geographers, for example, or even that of the Marxists silences the fact of disorder in the landscape. Space, rather, is, to paraphrase Massey, to be seen as the meeting up of different stories, juxtapositions of different power–knowledges through which new agents of a hybrid character can form: the result of a coming together in particular places of diverse influences at diverse geographic scales, creating new objects, which then become through their discursive understandings and their own power–knowledges and their spread elsewhere, transformative of geography: "Space/spatiality . . . is the sphere of the meeting up (or not) of multiple trajectories,[13] the sphere where they co-exist, affect each other, maybe come into conflict. . . . Subjects/objects are constructed through the space of those interrelations" (Massey, 1999a, p. 283).[14]

CASE STUDIES

The New Regional Geography

> What does make geography so difficult . . . is . . . its attempt to operate within specifically regional contexts. Ever since regional geography was declared to be dead . . . geographers, to their credit, have been trying to revivify it in some form or another. . . . We need to know more about the constitution of *regional* social formations, of *regional* articulations and *regional* transformations.
>
> —GREGORY (1978, p. 171)

> If places are indeed understood as bounded territories with internally generated authenticities, then localisms can indeed be problematical for the generation of wider understandings and wider movements (the problem of militant particularisms).

> But if the specificity of place is always understood as generated
> relationally then there is no simple divide between inside and
> outside, between local and global, between local struggles and
> wider movements. The question of the significance of "place"
> within geography was thus intimately related to the manner of its
> conceptualization.
>
> —MASSEY AND THRIFT (2003, p. 285)

The interest in relational space has been accompanied by a revival of interest in the character of particular places. The region had been central to human geography's classical phase in the first half of the century. It had then been marginalized by the spatial–quantitative revolution. The idiographic–nomothetic distinction drawn on in the debate about the possibility of laws in human geography condemned the region and places more generally to the status of a uniqueness beyond the possibility of explanation. As Gregory noted, what had continued was the use of places as contexts for research into causal configurations. Taaffe had noted this in his (1974) discussion of the spatial–quantitative revolution and little had changed by the end of the 1970s. The move to more relational concepts of space, however, opened up the possibility of a different approach to place, one in which the unique was not beyond explanation.

There are a range of possibilities here. More materialist approaches, based on ideas of what I have called the production of space, locate places and understand them with respect to wider processes but also what place-based agents in turn contribute to the changing character of the latter. The rejection of structures by postmodernism and poststructuralism opens up other sorts of explanation. I discuss these in turn.

Producing Places

The emphasis here is on the production of places, their differentiation from one another, in terms of more global structures of sociospatial relations, particularly those through which capital circulates. Accordingly, places can be identified in terms of their positions, or mix of positions, in wider spatial divisions of labor and consumption or in scalar divisions of labor. Massey's (1984) book on spatial divisions of labor provided an initial intimation of this pointing in the direction of branch plant towns or headquarters cities, but she was also keenly aware of the differentiating effects of the overlap of successive spatial divisions of labor, as in the copresence of mining and light assembly branch plant operations. The British locality studies project of the 1980s (Cooke, 1989; Harloe et al., 1990) followed closely in these footsteps, always contextualizing a locality in terms of changing positions in geographic divisions of labor or consumption. Power was clearly central to this vision. The emphasis in locality studies was on their competition for particular sorts of investment and so for enhanced niches in the geographic division of labor or simply their struggle to hold on to what they already had. This was by no means to argue that there was consensus within localities. Interests varied and were affected by the implications of new rounds of accumulation in very different ways as Massey herself showed in an article discussing the implications for class and gender relations of the superimposition of one geographic division of labor on another (1983).[15]

A crucial omission in the British locality studies project was addressing the question of why anyone would want to compete for investments in the first place. In urban studies in the United States there had been growing interest in the implications of fixed investments in place, in buildings, in land, for growth politics. Harvey Molotch's (1976) paper on "The City as a Growth Machine" was a landmark in this literature. The role of fixity in the politics of place was then developed in more ambitious fashion by David Harvey, as I indicated earlier in this chapter, in his geopolitics of capitalism. Through the contradiction between the fixity and the mobility of capital as it worked through different phases of its circulation, and the attempt on the part of those with fixed assets to negotiate it, territory and the defense of territory assumed an enhanced significance in the production of place. In Harvey's vision, growth coalitions formed at diverse geographic scales, often incorporating elements of the working class, and behind attempts to recapitalize the place, to restructure around new, growth sectors of the economy, and this often entailed social transformation: shifts in local educational provision and renegotiations of the social wage, of labor law, and of environmental legislation. Through their competition, people in places, orchestrated in particular by locally dependent capitals, produced not only themselves and the places they lived in, therefore, but other places too, albeit within the structural constraint that not everywhere could be high tech and rustbelt status would be visited on at least some.

Constructing Places

An alternative approach is to reject situating places and their transformation with respect to more global structures like capital, the state, and the division of labor, and recognize them as hybrid formations; as representing the coming together of different stories and discursive influences/knowledges, the geography behind their history, as Philo would put it. The best example of how this works out in practice comes not from a geographer but from Michael Peter Smith's (2001) *Transnational Urbanism*. Smith is a political scientist with an interest in the politics of particular cities but he has clearly been very influenced by the work of geographers. What Smith wants to emphasize is the irreducible particularity of cities and how this is conditioned by overlapping geographies of very different scope which intersect and juxtapose there. This is in contrast to what he regards as a reductive view of cities as vehicles of capital accumulation. Thus, according to Smith, the diverse racial and ethnic landscape of Los Angeles has less to do with a simple global logic of capital accumulation and more to do with a time-space juxtaposition of transnational migrants and refugees drawn to the city for other reasons. Mexicans originally came as agricultural workers but, subsequent to family reunification, low agricultural wages drove them into the city of Los Angeles, where they encountered, among others, upwardly mobile Korean professionals, pushed out of South Korea by government policies that produced more professionals than Korea's economy could absorb. The comings together then form the context for the conflicts emerging in particular places, conflicts that mobilize wider networks and through which places get constructed.

In this way Smith aims to disrupt the binaries through which, in his view, cities are typically understood: binaries of local and global, economic and cultural, universal and particular. People, firms, rather, are simultaneously local and global in their relations: hybrids, though he does not use that term. The distinction between the economic and the cultural, often mapped onto the global–local is a false one that is untrue to the way people live their lives. Rather, what one needs to do is recognize the significance of the networks that link local insides and global outsides, and transnational networks are a vehicle for doing this: "Situated actors socially construct historically specific projects that become localized within particular cities throughout the world, thereby shaping their urban politics and social life" (p. 60). Cities are sites where national and transnational practices become localized and from which local social practices reverberate transnationally, if not globally.

The structuralist arguments of Marxism and of world city narratives are, therefore, to be rejected. According to Smith, they represent positionalities with a power to construct the world in accord with particular interests (p. 9). His conception of transnational urbanism, on the other hand, is a way of deconstructing these binaries: of seeing insiders at the same time as outsiders; of deterritorialization requiring simultaneous territorialization; and of how the global as well as the local are constructed through "ordinary people" in particular places making use of their extended connections over space. The transnational urban is, therefore, the site of juxtapositions with the potential for transformative change.

Somewhat more ambivalent with respect to these sorts of arguments, but nevertheless highly sensitive to particularity and the effects of discourse, is a book by Allen, Massey, and Cochrane with the title *Rethinking the Region* (1998). There, the authors try to shed light on the emergence of what is now known, at least in Great Britain, as the Southeast: those parts of England that cluster around and include London, stretching south to the Channel, north into East Anglia and west toward Bristol. Over the period from the mid-1980s to the global crisis starting in 2008, this was the major growth region of the country. The label "the Southeast" or "London and the Southeast" acquired significant resonance among both its residents and those outside the area.

In reading Rethinking the Region, one is struck by a number of signposts, mutually reinforcing, indicating that this is a quite different approach from what might emerge from a more materialist production of place. Regional identities and their malleability are to the fore: how regions are part of wider systems of representation and of "otherness," such as "rustbelts," "poor regions," and "growth regions." This is argued as part of a discourse of regional dominance and subordination.[16] These regional identities get reworked over time. Their boundaries—always diffuse—shift, gaps open up within them, and what they are taken to stand for gets redefined. In the case of the Southeast, "growth" and a particular understanding of it as neoliberal has been an important ingredient. There has also been a move away from a simple contrast and complementarity between London and the rest of the region. This is a result of the emergence of new growth poles like Cambridge and Reading and the incorporation of older ones, like Southampton, and differentiation within the region between more exclusive residential areas and employment centers with a stronger working-class representation.

The Southeast is then taken to have been constructed on a particular coming together of events, conditions, and narratives. The area is defined by its status as a "successful" region that stands out from the rest of the country in terms of its growth. This growth can then be understood as the chance outcome in time and space of several different dynamics. Preconditions of longer standing were certainly important. They included *inter alia*, London as the center of national government and the major staging point for international connections, particularly in finance and airline connections. The fact that this was the area least affected by the heavy industry of the industrial revolution has also meant relatively scenic, "unspoiled" rural areas of the sort attractive to the technical managerial stratum keenly sought by high-tech industries. The contrast between an "industrial north" and a "prosperous commercial south" has been part of the popular British imaginary for at least a century.

It was within this context that the transformation of old regional growth sectors and the emergence of new ones laid the basis for a regionally distinct level of growth on the basis of which a new identity could be constructed. A key growth industry has been financial services. This has been important to the London economy for at least a century. It was, however, given a new burst of energy as a result of the government's reforms of the 1980s: notably the abolition of exchange controls and the reform of the Stock Exchange. Financial services have also become more common outside of London. With the rapid expansion of the industry and increasing housing costs in London, firms have established back offices in smaller towns surrounding the city, creating important spread effects. Quite independent of this and providing a further impetus to regional growth has been the expansion of telecommunications, computer services, and independent research and development in various centers. This owes something to the ability to attract the right sort of qualified labor in virtue of the attractiveness of the area as one to live in, but it is by no means the only thing. The chance presence in the Southeast of the two major research universities of Oxford and Cambridge has been important, particularly in the case of the latter, which became the nucleus of the country's equivalent of Silicon Valley. There were also earlier locations such as that of the pharmaceutical industry in Hertfordshire just to the north of London and of the aeronautical industry in Bristol at the western edge of the region.

There are other conditions that should be mentioned. Growth in the Southeast triggered a massive increase in housing values. Increasing incomes allowed people to bid more for the more attractive housing stock. But in addition limits on the supply of new housing entrenched in British planning legislation and the power of environmental lobbies have also played a major part. As housing values have risen, so has the ability to parlay that increased equity into borrowing, helping stimulate a consumer boom unmatched in the rest of the country and lending it a further distinctiveness.

Finally, it has been growth of a neoliberal kind that has put its stamp on the area. The emergence of the Southeast as a growth region occurred within the context of the neoliberal policies of Thatcherism. These were policies that in part targeted different parts of the region as in the reforms of the London stock exchange but also ones that affected the country as a whole but of which it was the principal beneficiary. The relaxation of the rules governing consumer credit interacted with the higher incomes of the region to give an additional impulse to its growth.

While I think that this book is the best example to date on the part of geographers of putting to work an understanding of the region that draws on elements of postmodern and poststructuralist thinking, it retains a stronger material base than is entirely consistent with the latter. It is by no means just narratives that are coming together in what is being constructed as the Southeast but real material events, like the growth of London's financial services, the spread of back offices into the rest of the Southeast, and the emergence of high tech. One can argue that these have been conditioned by a particular discourse of neoliberalism but the authors' focus does not lie exclusively there. There is also an awareness of material limits that would ring hollow in a more idealist reading truer to postmodernism and poststructuralism and that recalls some of the claims I made earlier when talking about the production of places. As the authors say, when discussing the Southeast and other regions in Great Britain, "Despite the political rhetoric of the 1980s, the south east could never have been a model for the rest of the UK: on the contrary, the relative decline of other regional economies was a necessary corollary to its growth" (p. 119). They then document the various modalities through which this occurred, including the movement of more skilled fractions of the workforce to the southeast at the expense of other regions.

The Politics of Scale

> Integral to the production of space, capital produces certain
> distinct spatial scales of social organization. These can be
> visualized as islands of absolute space in a sea of relative space. It
> will be necessary, then, to derive spatial scales out of the analysis
> of capitalist development and structure rather than simply to
> assume certain habitual scales as given.
>
> —N. SMITH (1984, p. 87)

Scale has always occupied a central place among geographical concepts. It has been used as an ordering device, allocating events and processes to local, regional, national, perhaps continental scales. More recently, under the aegis of relational concepts of space, it has acquired a much sharper analytic edge in which questions of power have come to predominate. One now talks not so much of scale as of "the politics of scale." There is nothing especially new about this. Mackinder and Bowman had a very similar view of the world. However, their contributions came within the first 30 years of the 20th century and were not argued out in abstract, indeed explicit fashion. Furthermore, they were followed by a period in which treatments were dominantly descriptive. Scale became simply a way of organizing material. Accordingly, one might recognize the significance of geographic difference at one scale, but other scales would take over at other points in the narrative. Agricultural variation might be understood in terms of underlying geology at a smaller scale and in terms of climatic variation at a larger one.

Spatial–quantitative geography had a more explicit approach to questions of scale and worked some change in how it was approached. In accord with its quantitative character, geographic variation was to be broken down into measurable scale components; one could now talk, for example, of *how much* variation in population density across American counties was associated with the counties themselves, with the states or with respective census regions, and, if one wanted, compare the results over time. This was an application of the variance decomposition approaches of the standard statistical tool of analysis of variance to data organized according to geographic scale.[17] One could learn, for example, what percent of the total variance was attributable to variation between the states, the counties, and so on (Moellering and Tobler, 1972).[18]

More importantly, though, breaking down geographic variation into scalar components was now allied to a relative concept of space. The processes working at different geographic scales were *spatial* processes. The principles of locational choice might be different at one scale than at another. Industrial firms might make a broad choice between the old Manufacturing Belt on the one hand and the rest of the United States on the other in terms of considerations of market access. But then at smaller geographical scales—within the Manufacturing Belt or within the rest of the United States—choice might be between cities with labor surpluses and those without.

What signaled a decisive end to this approach to scale and the advent of one based on a relational view of space was an article by Peter Taylor (1981) entitled "A Materialist Framework for Political Geography," even though at the time its significance might have gone unnoticed. This is partly because of the way he framed his discussion. This was less in terms of concepts of space than an attempt to move beyond work in political geography which had tended to hold events at different scales apart: so political geographers compartmentalized what happened at local, national, and more global levels, for example. Taylor wanted to show how events at different scales had to be seen as part of a broader whole in which interscalar relations were of cardinal significance. But he ended up by pointing to the way in which scale might be seen as an expression of social relations as well as showing how the scaling of social process might contribute to the reproduction of those social relations. In other words, his conception of scale was much closer to one based on a relational view of space.

Taylor's point of departure is Wallerstein's world systems theory, which in turn reflects the interest in Marxist geography initiated at the beginning of the 1970s. The choice of Wallerstein is highly significant since it represented recognition of the centrality of the global scale to what transpired at other scales, including that of the state, which had, up to that point, been so prominent in the work of political geographers. As I said, Taylor also recognized the way in which political geographers had tended to organize their writings around three particular geographic scales: the global, the national, and the urban. This is the basis for his suggestion that "the threefold arrangement has an important general function within modern capitalism" and that "We need a political economy of scale to unravel this situation." This is the background to his positing of relations between these scales starting with that of the world economy. This, he argues, is the scale of reality: "real" because this is the scale at which accumulation and, therefore, production is organized, reflecting the Marxist priority accorded to the latter. The urban, on the other hand, is the scale at which accumulation is experienced, since this is where people live their daily lives. So accumulation is experienced in the form of what people have to spend, whether or not they will be able to find work, pay the mortgage, and the like. Finally, the national is the scale of ideology. This is the scale at which interpretations of experience at the urban level are dispensed, in nationalistic forms, for example, all with a view to justifying subordination to the accumulation process, which is going on at the global scale.

This certainly represented a major step forward in how geographers viewed geographic scale. Scale and scalar relations were conditioned by social organization. But for a view inscribed in a thoroughly relational view of space, it was not quite there. This is because there is a functionalism about Taylor's argument that obscures the way in which scales and scalar relations are constructed: how they don't just come about because they happen to work but because people set about producing them and making them work; and they do this in virtue of particular sets of social relations like those defined by Wallerstein's world system. Here it is important to recall how Taylor accepted the tripartite definition of scales as appropriate to the study of political geography. He shows some doubt about the "naturalness" of these scales but then opts for them anyway. The achievement of those subscribing to more relational concepts of space was to call that naturalness into doubt. Scale, it now appeared, was something that should not be taken for granted; rather, it was something that was produced or socially constructed. Scalar concepts came, some even went, and this could be observed and understood. Categories like "the Rust Belt," "Silicon Valley," and "Southern California" had not always existed, and neither had the Wheat Belt, the Cotton Belt, or the Southeast (of England) (Allen et al., 1998).

Since the spatial–quantitative revolution, there have clearly been some quite radical changes in our understanding of scale. There are three that I would like to draw attention to here, though they are closely related to one another. The first is that scale is now seen as something that, as Smith avers in the quotation at the beginning of this section, is socially produced. One would concur with postmodernism and poststructuralism that it is a representation of the way the world is and that positionalities are involved in its production. But *pace* that understanding, not any representations will do. There is a material reality that means some work and some don't. Furthermore, we should note the way in which scale now came to be regarded as through-and-through political. Producing scale, whether national, regional, global, or whatever, involved the deployment of power; sometimes state power, sometimes not, and sometimes both and against other such deployments. Production was simultaneously contestation, therefore. Finally, production and contestation are typically of an interscalar sort. Production might be "bottom-up" or "top-down," suggesting a quite radical separation of scales. Initiatives for new scalar fixes might emerge at more central levels of the state, for example, or surge up from the grass roots through local political representatives or lobbying groups, though always with respect to social interests. In this way one could begin to make sense of new scalar forms of the state like the European Union or the World Trade Organization; or, alternatively the reorganization of corporate relations within a multinational around contested degrees of centralization or decentralization.

This conventional politics of scale has recently come under some critical attack. Prominent here have been papers by Amin (2002), Marston, Jones and Woodward (2005), and Allen both on his own (2010) and with Cochrane (2007, 2012). They show distinct overlaps with one another. They also seem to owe a lot in their understanding of the scale issue to Massey's work on how the construction of places is conditioned by the chance juxtaposition of a more extensive geography of relations. A major concern has been the hierarchical understanding of scale: how scales are ordered in subordinate–dominant relations as in that between the local and the global or between local government and central government. A related one has been the territorial emphasis of scalar thinking. The power of the network is the central consideration in destabilizing notions of scale and territory.

Scalar thinking, it is argued, is to emphasize vertical relations. The image of nested agencies of the state at different levels with those at higher levels constraining, even determining, what those at lower levels do is a common one. This is contrasted with the horizontality of networks, which interfere with top-down determination, opening up the possibility of resolutions of a more contingent, open character than those imposed from above. Verticality also implies a very centered view of the world: The upper levels of the state are where the important things happen and decisions are then dispersed to lower levels. Networks, on the other hand, are seen as a vehicle for decentralization of decision making, and more democratic therefore. Similar arguments are made in the case of globalization and *its* politics of scale. Globalization is not necessarily the all-determining frame of life at whatever subordinate scale one wants to consider it. Again, there is a contingency to outcomes, a contingency mediated by relations of a more horizontal sort.

As far as the territorial frame of scalar thinking is concerned, networks, implicitly defined as horizontal, are seen as an agent of deterritorialization, as facilitating the dissolution of local attachments, the intimate local connections that firms often enter into and other forms of local or place dependence (Amin, 2002). Territorial action, exclusion, inclusion, locally focused forms of development, therefore, dissolve away as dependence on the local, material and affective is loosened. Firms become multinational; labor markets, at least for some, become continental, even of global proportions. Castells's space of flows (1983) is the most reductive of these arguments. In turn, they have led to new claims about the relation between space and place: how place as a horizon of action in people's lives is being undermined by space.

Much of this critical literature has drawn impetus from poststructural and postmodern forms of thinking, and this, as such, represents a rejection of the sort of political economy embraced by historical materialists. Scalar thinking is now seen as a particular form of discourse: a discourse for some and to the disadvantage of many; as a positioning for some sort of leverage. It is not, in other words, the way in which capitalist states, in virtue of the limits and possibilities of the capitalist space economy, *have* to be organized. So too is it with territory. As Allen and Cochrane remark, "It would seem that the language of territorial politics is not only stubborn, but equally that it cannot simply be wished away by some conceptual wand, since it is itself a powerful political construction. Assemblies, regional development agencies, and the like, are performed as territorial entities that try to hold down the fluid elements of global life in the general interest of their 'regions'—seeking to generate fixity through 'processes of government and governance'" (2007, pp. 1162–1163).[19]

Boundaries

The study of boundaries has a long history in political geography. Here I want to confine myself to their international expression. Historically much of the work here drew on ideas of absolute space. International boundaries were lines on maps that might be correlated with other features. They might demarcate not just distinct states but *nation* states; in other words, they separated those with different national identity. A dominant interpretive slant was boundaries as "natural": the degree to which they followed natural features of the landscape like mountain ranges or major rivers.

With the turn to the spatial–quantitative there was a clear shift in emphasis. Boundaries then became an object of interest because of their implications for interaction over space. An early expression of this was a study by Ross Mackay (1958) examining telephone call activity between different cities in Canada and the United States. Using a gravity model he showed that, holding constant its distance and population terms, the number of calls between cities in Canada and the United States fell compared with traffic between similar pairs of cities in Canada.[20,21]

This sort of research, perhaps in this instance banal in its conclusions, was entirely in accord with the priorities of the spatial–quantitative revolution and the search for spatial order. Drawing on the fundamental spatial properties that would define relative space, considerable progress could be made in demonstrating spatial regularity in human activities, including boundary effects on telephone call volumes. The subsequent embrace of concepts of space that were more relational in character would change this. Now it was not so much the effects of boundaries on movement and other distributions that became significant as their production as part of wider social—more accurately socio*spatial*—processes. Some of the possibilities were grasped in an article by Newman and Paasi (1998). Their agenda is clear:

> State boundaries are equally social, political and discursive constructs, not just static naturalized categories located between states. Boundaries and their meanings are historically contingent, and they are part of the production and institutionalization of territories and territoriality. (p. 187)

Attention now shifted to what they call "boundary *producing* practices" (emphasis added) and their social construction.

An excellent example of this more recent approach to boundaries is a 2007 article by Mat Coleman, in which the author identifies a shift in the geographic scope of the arena within which boundary effects are produced. This is a state-induced shift away from the boundary itself, where passports and visas are typically checked, to wider spheres. These include points that are within the space being bounded as well as spaces beyond: a process of transfer that he refers to as "important changes in the spatiality of state geopolitical practice in the borderlands" (p. 609). His particular case is that of the United States–Mexican border. This has been the site of contradictory forces. On the one hand, there have been pressures to facilitate the movement of goods between the two countries and to limit delays at the border. On the other hand, heightened security concerns subsequent to the attacks on the World Trade Center in New York in 2001 have resulted in a case for intensified monitoring that is in turn likely to cause delays. The contradiction has been negotiated by the federal authorities in two ways. On the one hand, there has been increased provision for the apprehension of undocumented immigrants within the United States. The federal Department of Homeland Security has deputized state and local police officials to detain and to request deportation. On the other, federal power has been extended beyond the borders of the United States in the form of collaborative action with authorities in

Central America designed to apprehend the undocumented before they even reach the border:

> For example, Operation Global Reach—a Department of Justice project head-quartered in a new Immigration and Naturalization Service (INS) regional office in Mexico City—established a permanent US immigration policing presence throughout Mexico and Central America in 1997. The object of Operation Global Reach was to find ways to stop undocumented migrants and arrest smugglers *before* their arrival at US borders. This took the form of joint enforcement operations with host country law enforcement officials (primarily Mexico) as well as through training non-US immigration authori-ties to act independently as US immigration enforcement proxies. (p. 620)

This clearly signifies a rather different agenda from that of gravity model studies where "boundary" is a dummy variable or studies like those of Taylor (1971), where it is other locational effects of boundaries that are in question. Rather, attention now shifts to the social construction of the boundaries themselves.[22] Even so, there is a complementarity between these two sorts of study that is worth noting. On the one hand, one might argue that the production of boundaries presupposes the sorts of inter-action patterns studied under the rubric of relative space. Among other things, the emergence of some national homogeneity and identity encour-ages some differentiation in the geography of people's interaction over space. On the other hand, the further sedimentation of boundaries and their barrier effects further entrenches those interaction patterns. It is as if some proto-national feeling that generates particular patterns of socio*spatial* interaction is a necessary precondition for boundary produc-tion. Acts of boundary production then reproduce and transform—in this instance "intensify"—the original precondition.

CONCLUDING COMMENTS

Space, or more precisely space relations, is geography's key concept. But at the beginning of the last century, consideration of what it meant, of how it was drawn on in practice, both academic and lay, how ideas about it were socially structured, was almost entirely absent. Only with the spa-tial–quantitative revolution, that great hinge of 20th-century geography, did this begin to break down, albeit, as with so much about that particu-lar initiative, more implicitly than otherwise. It was a concept of relative space that was to be the vehicle for achieving the goal of making geogra-phy a generalizing science and so, it was hoped, a respectable academic discipline.

Fundamental to spatial–quantitative geography was the recognition that location could be defined not necessarily in terms of some imaginary grid or a framework like that provided by a map of countries with respect to which one could locate cases. Rather, it could be defined in terms of particular objects as in distance from, accessibility to, or direction with respect to a city, a line of transportation, a source of raw materials, or whatever. Significantly, we should recall that the history of a conception of space as relative in human geography is quite a bit older than the spatial–quantitative revolution. Human geographers had always been interested in accessibility as in Mark Jefferson's buffer zones,[23] Mackinder's strategic geography, or Vaughan Cornish's (1923) capital cities. They had also been interested in movement and origin–destination relations, as in Sauer's work (1952) on the spread of domesticated plants.

The genius of the spatial–quantitative revolution, though, was to hitch a concept of relative space to the challenges and opportunities of a particular conjuncture in the history of human geography. The challenge was the weak status of geography as a university discipline in the 1940s and 1950s. The embrace of relative space promised a renewal of human geography around the conception of spatial regularity and hence its establishment as a law-seeking discipline; in other words, as "scientific." The second conjunctural feature was the emergence of a set of material problems around location in urban and regional planning: notably, how to plan for new highways and retail centers in a period of rapid metropolitan expansion. A scientifically engaged human geography drawing on conceptions of relative space and the sort of order provided by the gravity model or linear programming's solutions to the transportation problem offered answers.

From the standpoint of the intellectual coherence of human geography, though, exploiting this particular opening came at a cost. As human geography began to see itself as increasingly a social science, with the emphasis on "science," it looked around for peer disciplines. Economics with its emphasis on measurement and a similar passion for scientific credentials would be the peer discipline of choice. Location theory then secured the connection. This in turn meant a commitment to short-term equilibrium processes whereby changing locations would be seen as adjustments to external shocks.[24] Given the way in which the pure competition of neoclassical economics was supposed to eliminate exploitation (see Chapter 3) it also meant the elimination of power relations from the new geography, something not necessarily entailed by conceptions of space as relative, as Mackinder had made clear in his earlier work. This came together in a seemingly harmonious manner with the technocratic processes of planning. Planning practice was reactive, adjusting to trends and so part of the process of achieving a new spatial equilibrium through facilitating the location of new highways and new retail centers.

These forces then combined to induce a disinterest in areal

differentiation, in places and in their content. The idea that geography could be a law-seeking field centered on locational regularities precluded anything to do with the particular: places or regions were "particular" and resistant to explanation along "scientific" lines. It is significant that a central primer of the spatial–quantitative revolution, Peter Haggett's *Locational Analysis in Human Geography*, failed to include area in its categorization of the elements of regional analysis (1965, p. 18). Haggett's focus was famously on movements; networks; nodes; hierarchies; and surfaces, with the latter conceived as a density map of points per unit area. The central problematic of the spatial–quantitative revolution was location, and location was to be understood in terms of relative space.

But in fact a concept of space as absolute was never totally displaced, nor could it have been. In his Presidential Address of 1974 on "The Spatial View in Context," Taaffe acknowledged the significance of region as context: as a space within which spatial relations of a quantifiable sort might be different from elsewhere. And once spatial interaction breached national boundaries as in Mackay's work on interaction effects or was configured by Peter Taylor's (1971) national territorial shapes, one was once again back in the realm of absolute space, even if this went unrecognized. Similar comments can be made about the identification of scale components of variance. The spatial analysts claimed that it provided clues regarding the ways in which particular spatial processes were scale specific but the way in which the analyses were framed by the absolute spaces of counties and other territorial units went unremarked.

It would be wrong, though, to say that it was contradictions of this sort that led to the supersession of relative space as the dominant understanding in human geography. Rather, the emergence of relational space had much more to do with underlying assumptions about social relations as they played themselves out over space. As we saw in Chapter 3, spatial–quantitative geographers had some ideas of their own about social process. But in their neglect of power relations and their belief in the rule of an impartial knowledge, their ideas could not meet the challenge of urban and regional questions, and later the ecological question, as they were increasingly perceived from the early 1970s on. The void was filled by a more critical form of social theory: first historical geographical materialism and then the "posts." These necessarily entailed new conceptions of space: Space would henceforth be grasped as an essential aspect of the social process.

Parenthetically I should note how utterly compatible the spatial–quantitative revolution's conception of relative space was with its appropriation, through location theory, of the tenets of neo-classical economics. As Marx tirelessly emphasized, the conceptions on which classical political economy relied were fetishizations: individuals without social relations, social relations without individuals, capital without social conditions—a world of things without necessary presuppositions outside of themselves, therefore. These would be its legacy for neoclassical economics. Space or spatial arrangement was viewed in exactly the same way: as an external condition for social life, generating notions of spatial fetishism; or as something into which the planning apparatus could intervene without fear of unintended consequences. This was, of course, as Marx again explained, how the bourgeois world presented itself, how it was experienced.

A conception of space as relational, on the other hand, recognizes that the objects with respect to which relative space is defined have relations one with another that are of a necessary sort; that they are what they are in virtue of their relations with other objects and conditions. Space relations are seen as a necessary aspect of the social process. The gravity model was a favored device for the spatial–quantitative geographers. Spatial interaction was a function of the relative sizes of the cities interacting and the distance separating them. What relations cities internalized in order to make people there *want* to interact with those elsewhere went unremarked. Yet clearly the gravity model presupposed a whole set of social relations: relations not just between cities but conditioning urbanization as a whole, notably those relations of production that set in motion processes of exchange across space but also those that created such objects of analysis for the gravity model as shoppers and commuters.

Recognition of the necessary relation between space and social process provided a new slant on absolute and relative space. Rather than being superseded, they were understood in a new light: as definitions that made sense for the particular practical purposes entailed by the social process. Rent could be conceived from the standpoint of absolute, relative, and relational space. At any one time a property owner might be concerned with the limits of her property or with absolute space; with the varying accessibilities contributing to its value, or relative space; or, yet again, considerations of space as relational might surge to the fore if the social status of rent came into question, as indeed it has when, for example, the so-called "takings clause" has become an issue in the United States.

NOTES

1. Though implicitly, of course, this intervention into spatial arrangement denied its all-determining qualities, its thing-like existence outside of any presuppositions.
2. Thrift (1994, p. 216) blamed the Marxist legacy for this: "The neglect of regional differences in Marx made it difficult for Marxist geographers, as they appeared on the Anglo-American scene in the 1960s and 1970s, to talk

about geography. Certainly, too many of the early attempts by the new Marxist geographies ended up looking like crude forms of economic determinism."

3. Compare Thrift (1994, p. 208): "Universal forces have to be lived by people,
   and they can often be put together in all kinds of unexpected ways. These
   different ways of living, sedimented over the ages, contain their own powers to react back on universal forces, changing them, even producing new
   forces."

4. The qualifications were first, that in increasing production the division of
   labor went hand in hand with providing workers with improved means of
   production; and that this in turn presupposed a particular set of social relations of production.

5. See Sayer and Walker (1992, pp. 152–153).

6. As he points out, the technonatural revolution was an opportunity for
   cementing networks of relations at even larger geographical scales. In its initial stages under Franco water development was desperately short of capital.
   A way around this was to exploit the Cold War and trade military bases to
   the United States in exchange for aid that could be devoted to it: something
   that the Americans could support anyway because of its promise of securing
   a stable Spain that would be "on the right side" in the Cold War.

7. For example, she seems to have little patience for the excesses of antifoun-
   dationalism: "Geographical imaginations (for instance of regions and of
   regional uneven development) are not simply mirrors; they are *in some sense*
   constitutive figurations; *in some sense* they 'produce' the world in which we
   live and within which they are themselves constructed. On the other hand,
   the very fact of the attempt at intervention implies a rejection of that position which would entrap us in a prison house of language to which there is
   no outside with any force whatsoever. (In other words, I am loathe to turn up
   on the steps of Downing Street in order to offer another metaphor/story/
   imagination whose relationship to 'the extradiscursive reality' of the North–
   South divide is either nonexistent or totally unknown)" (2001, p. 10).

8. Her work shows the consecutive influence of many different layers: another
   instance of the geological metaphor she turned to such advantage in talking about regional specificity (1983) and the interaction of earlier with later
   "deposits." There was an earlier involvement with Marxism followed by a
   turn to what seemed to be critical realism before people were talking about
   it and then more recently an engagement with poststructural and postcolonial ideas that has been as stimulating as anything she has done in a career
   characterized by high imagination. And this has always been against the
   background of a concern for equality of both treatment and outcome and
   social justice.

9. Though note Thrift's earlier intimations of this in his contribution to the
   debate about Harvey's work in *Society and Space* in 1987 and with reference
   to the production of subjects in local contexts: "These local contexts provide
   all manner of unexpected combinations of "universal" social relations which
   can, in time, produce subjects with hitherto unsuspected or unconsidered
   capacities to act, either singly or more likely in league with others, to modify
   existing social relations or even forge new ones" (1987, p. 403).

10. As Gregory (1990) pointed out, while a historicizing dimension had, to some

degree, at least, characterized modernity, that did not necessarily mean that a spatial turn in the social sciences was postmodern.

11. For a different take on this issue, but one which also emphasizes the need to spatialize time-dependent understandings of geography, see Taylor (1989).

12. W. W. Rostow's (1960) *Stages of Economic Growth* is, in this regard, a classic modernist statement.

13. See Gidwani and Sivaramakrishnan (2003) for a nice drawing out of the implications of this "meeting up of multiple trajectories."

14. Compare Thrift: "The account of place I have offered in this chapter is . . . one based in a *relational materialism* which depends upon conceiving of the world as associational, as an imbroglio of heterogeneous and more or less expansive hybrids performing 'not one but many worlds' and weaving all manner of spaces and times as they do so" (1999, p. 317).

15. There are some excellent contributions from sociologists on this sort of superimposition. See Warde (1988) and Gendron and Domhoff (2008).

16. "As part of an imaginary regional geography of the United Kingdom, the south east holds pole position in a *discourse of dominance*" (p. 10).

17. Though one should note that the relationship of jurisdictional units organized in some sort of hierarchy to simple notions of physical scale is, at best, variable.

18. This doesn't sound particularly fruitful, but the approach could be when allied to some sort of theory or claim: to what extent, for example, has the division of variation in American partisan choice between congressional districts versus states tended to change over time? In other words: are people now taking their voting cues more from senatorial races than from those for the House than they might have done in the past?

19. For a critical consideration of these arguments, see Cox (2013b).

20. Similar effects were reported for telephone calls within and between the two Australian states of Victoria and South Australia; see Logan (1968). For boundary effects in general from a viewpoint within spatial–quantitative geography, see Cox (1972a, Chap. 7).

21. Much more ambitious and altogether striking in its originality was an attempt by Peter Taylor (1971) to demonstrate the effects of territorial shape (i.e., the shape defined by a state's boundaries) on interaction *within* that shape rather than, as in Mackay's case, interaction *across* boundaries.

22. See also Juliet Fall's excellent critique of the work of economists drawing on ideas about boundaries (2010) and their reified views of them.

23. See *www.csiss.org/classics/content/12* (retrieved June 30, 2013).

24. As in the concept of weak competition critiqued by Storper and Walker (1989, Chap. 2).

CHAPTER 6

Methods in Question

In raising the curtain on human geography's growing embrace of critical social theory, historical geographical materialism and humanistic geography led the way. They were both positioned in very different understandings of the world than had been the case of the spatial–quantitative revolution. They could not, therefore, be merely critiques of the way in which spatial–quantitative work had drawn on concepts of the social. They were also thoroughgoing critiques at the level of method. Both rejected the positivism of the spatial–quantitative geographers but in very distinct and contrasting ways.

The spatial–quantitative revolution itself that had originally started the debate about method and made geographers think about it in ways that they previously never had. But in retrospect the new understanding of method was very limited and necessarily so.[1] The focus was on technique in the narrow sense: what statistical tests or operations to use where and when, the development of new algorithms for identifying solutions, the meaning of the coefficients subsequently calculated, and the pitfalls of certain modes of statistical inference. The effect of the new initiatives of Marxist and humanistic provenance was a complete reordering of priorities. There was a new, broader sense of method: method as more than that implied by a kit bag of tools to be selected from as the concrete research tasks demanded. Method could now be conceived first and foremost in a conceptual sense. The attack on positivism demanded that questions of the relation between facts, values, and theories be pushed to the forefront. This then widened out, particularly under the influence of critical realism, to include questions of different modes of theorizing and of abstraction, the limits and possibilities of explanation as opposed to interpretation, and the possibility of truth.

In the second place, it has seemed that theory and technique changed position. Under the spatial–quantitative revolution, technique assumed pride of place, and theory, while certainly apparent and a major point of reference, was taken for granted. One of its distinctive features had been the interest in technique apart from particular empirical applications. There was an abundance of articles with titles like "A Method for the Geographical Analysis of X," "A Family of Probability Distributions for Y," or "Geographical Interpretations of W Values," along with numerous textbooks demonstrating "how to do it" (Cole and King, 1968; Haggett, 1965; King, 1969; Norcliffe, 1977; Yeates, 1974). With the growing interest in social theory of a more critical sort, from the early 1970s on there was little or no equivalent of this unless it is Sayer's (1984) very different book on *Method in Social Science*.[2] This shift in emphasis is significant. Spatial–quantitative geography had theory and was self-conscious about it: It was something that its practitioners believed differentiated it from the field's "prescientific" past. But as Trevor Barnes (2001) has emphasized, it was theory that could be taken for granted and not debated. The major research issues were not seen as theoretical. Nor were they conceptual. Explanation, cause, and truth were not to be argued about. Rather, it was a matter of improved measurement: the need to spatialize statistics, to develop better indices of dispersion, segregation, and the like and of ways of identifying growth poles. Once the facts had been ascertained in this way, then theory could be added to in inductive fashion. It wasn't just that facts were seen as separate from theory; the development of theory and its revision depended on the accumulation of empirical regularities and their acceptance as reliable generalizations.[3] The self-conscious engagement with social theory led to a dramatic shift in priorities. Contesting theoretical claims, claims about the nature of the world, now took center stage. Theory would lead method and not the other way around.

For the "posts," the positivism that was so clearly in the line of sight of humanistic geography, and to a lesser degree of historical geographical materialism, takes a back seat. Notably, the spatial–quantitative revolution is no longer an issue. This is a battle long since won. Positivism now becomes one aspect, if an integral one, of grand theory, particularly that of modernization. The idea of cumulative, objective knowledge as prefatory to human emancipation was clearly one that would sit ill with those skeptical of any universal claims to knowledge. Rather, science was now interpreted as a local conceptual practice bearing the mark of the sociohistorical embeddedness of its producers. Science's claim to epistemic privilege was seen as a tactic to exclude, silence, and disempower, and indeed there was plenty of evidence to encourage that claim.

Methodological debate in human geography has, therefore, been intense and noisy. There are numerous ways in which one might slice through these debates, though necessarily the slices will overlap. In what follows, therefore, they should be taken as particular lenses through which to examine the issues of methodology that have been so much to the forefront since the spatial–quantitative revolution and in ways that they never were. An obvious starting point is to examine the quantitative–qualitative couplet. After that, any logical order becomes less clear. In any case, reference to earlier discussions is necessary. In no particular order, therefore, but taking them in turn in a quite arbitrary way, I examine three different sets of contrasting approaches: analytic versus interpretive; categorial versus dialectical; and pluralizing versus totalizing. These quite broad categories will allow me to discuss the wide range of issues that have indeed surged to the fore at various times without, I hope, forcing them into procrustean beds.

QUANTITATIVE–QUALITATIVE

Quantitative geography was positivist. But it was almost after it was over that it was recognized that this might be a problem. It was the critics who pointed to its positivist credentials. It seems in retrospect that the quantitative geographers had been too busy assembling data sets and testing hypotheses to notice. Now, as spatial–quantitative geography came under more critical scrutiny, it was its more fundamental assumptions in the philosophy of science that were called into question.

Observation was privileged. It was only after the data had been brought together and hypothesized relations tested and verified that one could begin to make claims of a theoretical sort. Correlation in the statistical sense was a crucial weapon. Only by identifying constant conjunctions of events, this time over space, could one make claims about causation; causation was sequence. Theory was about events and not about any unobservable relations that might account, or provide necessary conditions, for those correlations. Likewise, whether a particular geographic outcome was good or bad was something beyond the scientific geographer's purview, though the interest in efficiency solutions to location problems suggested some slippage here.

The fit between positivism and quantitative methods was of an almost organic sort. Quantitative methods seemed to provide a value-neutral approach to observation; a magnitude was neither a good nor a bad thing. It also provided objectivity. Measurement was an impeccable check on the observations of others. And there again, what better way for evaluating correlations, than holding variables constant to detect constant conjunctions of events that might otherwise have gone undetected?

The attack on positivism, therefore, had to be an attack on quantitative methods. This was to be one from which quantitative geography has had difficulty recovering: a retreat into its own closet rather than an attempt to respond in the terms of the critics and show how quantitative methods nevertheless remain a crucial element in the methods appropriate to human geography. This was, and remains, unfortunate. There is no reason in principle why theoretically informed studies should not employ quantitative analyses of data either as a starting point—what needs to be explained—or as corroboration for an explanation. There are undoubtedly some issues in the use of quantitative methods. But on the whole they remain underused by those who define themselves as critical human geographers or Marxist geographers. This has been a serious loss to the field and continues despite ongoing calls for the use of "mixed methods."

As I said, the advantages of quantitative methods seemed very clear at the time. This was particularly so in comparison with what hitherto had passed for method in human geography. They promised an objectivity that had necessarily eluded those who had been content with map comparisons. Now one could determine the degree of association through a correlation coefficient and dispense with "eyeballing." Degrees of accessibility could be placed on a finer spectrum of variation than one differentiating simply between "more" or "less." Drawing upon the same data, someone else could do the same thing, removing the process of establishing associations from the biases of a particular observer. This remains the case. Even so, and as became increasingly clear with practice and with criticisms from outside the spatial–quantitative revolution, quantification of geographic relations poses challenges.

The measurement of events, their degree, their frequency, and their relative locations might in some cases appear straightforward: The populations of cities or the actual distance between two points measured from a map provide examples. In fact, though, it quickly gets one into issues of what is a city and what does distance mean. Again, one might be able to improve one's measurements: taking travel time instead of distance between two cities. For something like a city, though, conventions have to be adopted, and they are typically the conventions of a national data-gathering agency, like a census office. Accordingly, political definitions might be used, taking municipalities as cities. This again, though, is highly problematic. It is not just that the criteria for municipal status might be different between countries, which is certainly a problem for international comparisons; it is also a matter of translating, for example, a jurisdictionally highly fragmented landscape into the urban regions that one wants to study.

Exactly what a number means can also be in question. The meanings of some data are relatively transparent, as in many demographic or economic data: something like life expectancy or productivity per worker, though one has to qualify even that by reference to the very substantial variations within countries. Other data are more refractory. In studies of the geography of voting, determining the relation between partisan choice and social class may be the objective in view; not all national censuses collect data on social class but the British one does. What, however, does a vote for the Conservative Party or belonging to one of the particular social class categories mean to the people doing the actual voting? This can vary a great deal, and often by region and their disparate histories of class relations. Even something as seemingly harmless as the study of home values in a metropolitan area can encounter problems of meaning; living in a desirable school district does not carry the same weight for everyone.

A further point is that many of the objects we encounter in social science are already quantified (e.g., census statistics, "class" in the British Census, prices, rents, wages, crime statistics, housing starts, local government expenditures, taxes, migrants, refugees). This means that we have to ask, for whom are these statistics useful and how does that affect their definition? They may be useful for the purposes of the people collecting them but not for social scientists. Often the sorts of qualitative variations we might be interested in as social scientists are overlooked: What is really a petty crime may be counted along with serious crimes in order to pad the figures prior to a funding request by the Police Department! Whether pupil–teacher ratios provide useful comment on inequalities depends on the qualitative characteristics of the teachers (and of the pupils as well!).

Quantitative analysis has had strong explanatory goals. Correlations are calculated and regressions are run with a view to establishing what is related to what and how that might shed light on causal relations. But again there are some serious issues here that make the task more daunting than it might initially appear. One of the difficulties, as Andrew Sayer (1992, p. 179) emphasized, is that quantitative methods are very unhelpful when it comes to identifying causal conditions or relations. The expression of relations in quantitative terms—$Y = f(X)$ or $Y = X1 + X2 + X3$—tells us *nothing* about cause and effect. The equals sign connotes nothing in and of itself about what produces what; the two sides of these expressions could be interchanged without any loss of meaning. Even if the variables are given some ordering in terms of time—for example, $T2 = f(T1)$—mathematics cannot confirm a causal relation, unless, of course, we have reasons *of a conceptual nature* suggesting that $T1$ is a necessary precondition for $T2$.[4] So mathematics is no substitute for careful theorization.

A second problem aside from the equal sign is that one is dealing solely with quantities. As Sayer again pointed out, these cannot tell us anything about the causal properties in play: They merely tell us the magnitude of the effects or the number of times they occur, when activated. In other words, and as Massey averred many years ago (1984, p. 12), spatial variation in something does not explain spatial variation in something else. During the height of the interest generated by the relative shift of manufacturing employment in the United States from the so-called Cold Belt to the Sun Belt, observers were fond of pointing out the relation between employment growth across states and various measures of "business climate": in particular, weak union membership and labor laws that made it more difficult to organize. What was making the shift possible, though—what was empowering capitalists to shift locations, notably the deskilling of various labor processes—went unremarked.

It was within this context that qualitative methods seemed to offer a more promising approach to unearthing processes. These have always covered a huge variety of methods, ranging from the relatively unstructured interview in which the interviewee is encouraged to take the initiative with a view to revealing the unexpected but the pertinent; through life histories and ethnographies; to the analysis of documents, often referred to as archival analysis, and even examination of company strategies appearing in the financial press. Through these techniques, it was believed that meanings could be probed, reasons for actions inferred, and causal structures identified.[5] With respect to the latter, life histories of people sharing some particular status or occupation might reveal a repetitive pattern of conditions that both limited and facilitated particular career paths (Bertaux, 1981). A major emphasis has been on the interpretive involving a to-and-fro between the investigator's framework of meaning and that of the people being studied: the so-called hermeneutic or process of "verstehen," discussed in detail in the next section.

A crucial maneuver is contextualizing or situating action. Cole Harris (1978) put this at the heart of what he called "the historical mind": the accumulation of all manner of data about a period/place and the attempt to put them in the context of what is known about that period/place. With the accumulation of knowledge, the context gets adjusted, allowing reinterpretation of other historical materials, autobiography, and textual materials like novels and newspaper accounts (see also Thompson, 1971). This serves to underline a major contrast with quantitative methods as they have been traditionally practiced, since the tendency there has been to *de*contextualize. The assumption in inferential statistics that observations are "independent" of one another is only the most extreme expression of this, though the discovery and development of studies of spatial autocorrelation has been mitigating.

In an important sense, qualitative methods are subversive. In terms of their data sources, quantitative methods represent a top-down view of the world: a view through the lenses of those who design questionnaires and censuses, a view that smoothes out and necessarily homogenizes. Qualitative methods are an attempt to understand the actions of people by talking to those who do the acting. This does not mean that questions of power evaporate. On the contrary, important questions of positionality remain whether it is a matter of interacting with people or interpreting the archives.[6]

With these considerations in mind, there has been a tendency to see quantitative and qualitative methods in sharply contrasting terms: descriptive–explanatory, decontextualizing–contextualizing, brute data–meanings. This was the position of Sayer (1984) in his highly influential *Method in Social Science*. He makes several points that, in themselves, seem undeniable and that led to his distinction between what he calls intensive and extensive methods: an observation that has resonated widely in human geography and encouraged interest in what have come to be known as "mixed methods." "Intensive" research is focused primarily on causal structures and secondarily on the contingent conditions operating in particular instances. Knowledge here was to be gained through in-depth interviews, along with focus groups and participant observation.[7] It could also work through a close reading of the work of others along with thought experiments in order to develop a structural analysis appropriate to the case under consideration: that is, what are the conditions that have to be present in order for X to have the particular causal properties that it does? In intensive research, one worked from reasons to conditions to the structures that enabled and limited and then to the contingent conditions operative in the particular case under study. Causal research, therefore, depended on the examination of particular cases of some phenomenon: neighborhood activism, gentrification, the reorganization of firm divisions of labor or of firms in a particular sector of the economy. This gave it a sharply contextualizing character.

For Sayer, however, this did not mean a rejection of more quantitative forms of analysis or what he called "extensive" research. Certainly, these could not result in the identification of causal mechanisms. In the instance of any empirical regularity, these would typically be quite diverse. Quantitative, generalizing forms of analysis pulled observations out of contexts characterized by distinctive structures of social relations and contingent conditions. There were indeed important differences between the two research modes (see Table 6.1). On the other hand, there were, as Sayer noted, complementarities between them. He did not go into detail on what those complementarities might be. In retrospect, this was unfortunate. This is because it added to the marginalization of quantitative methods that was already underway in human geography and so cut it off from a way of thinking and a set of tools that can be immensely important at particular stages of the research process.[8]

Not least, research into empirical regularities, whether those regularities are in the form of correlations, trends over time, segregation or other indices of localization, can help give direction to more intensive forms of research. Alternatively, the identification of residuals from regression can open the way for contextualizing forms of analysis aimed at identifying

TABLE 6.1. Intensive and Extensive Methods: A Summary

	Intensive methods	Extensive methods
Research question	How does a process work in a particular case or number of cases? What produces a certain change? What did the agents actually do?	What are the regularities, distinguishing features of a population?
Type of account produced	Causal explanation of the production of certain objects or events, though not necessarily representative ones	Descriptive "representative" generalizations, lacking in explanatory penetration
Typical methods	Study of individual agents in their causal contexts, interactive interviews, ethnography	Statistical analysis. Large-scale survey of population or representative sample, formal questionnaires, standardized interviews
Appropriate tests	Corroboration	Replication

Note. Based on Figure 13 (p. 243) in Sayer (1992). Copyright 1992 by Andrew Sayer. Adapted by permission.

just why it is that people in a particular place tend to be, for example, more left wing in their voting than its social composition might suggest. Having established some regularity or departures from it, intensive methods then help shed light on just why that is. This was the approach so effectively adopted by Kim England (1993) in her research into what she called "suburban pink collar ghettoes."

Even so, the distinctions identified in contrasting intensive and extensive methods thus were always overdrawn. Quantitative methods research could do more than identify "descriptive 'representative' generalizations, lacking in explanatory penetration." Rather, some spatial–quantitative work *has* been concerned with uncovering mechanisms of the sort that Sayer believes are only accessible through "intensive methods." The work of Hägerstrand (1957, 1967) is notable. Through the Monte Carlo simulation of the diffusion of innovation and of migration, he sought corroboration for his hypotheses as to the exact nature of the different processes. Monte Carlo simulation worked through a specification of the probabilities of certain events occurring. These probabilities incorporated speculations as to the nature of the underlying process. To the extent that the simulated distributions approximated in their general form to observed ones, then at least the particular mechanism under study could not be rejected.[9]

Likewise, much depends on the degree to which the quantitative analysis is informed by a careful conceptualization of the relations, including causal ones that are being evaluated. There was indeed heavy emphasis on empirical regularities on the part of many of the practitioners of spatial–quantitative geography, but this wasn't necessarily with a view to providing the essential key to explanation. Care was typically taken in moving from regularity to an explanatory account. Correlations would be only one part of the basis for arriving at an explanation for some geographical distribution. Case studies were commonly drawn on in the attempt to give a study some explanatory weight. In other instances, location theory might provide the necessary assistance in interpreting a set of correlations or regressions. One can sneer at the resort to "hypothesis testing" but what was being evaluated quantitatively typically had some basis in theory, sociological, economic, or otherwise. Papers invariably included a literature review, and this was typically more than a recounting of the degrees of relationship identified in previous quantitative studies.

ANALYTIC–INTERPRETIVE

The dualism of analysis and interpretation made its way initially into human geography by way of the priorities of humanistic geography. The central issue was that of the significance to be accorded to meaning when understanding human geographies, in particular the meanings ascribed to the world by those acting in it. The view was that first of all, this was crucial to any defensible explanation; in this regard, it was following the protocols of interpretive social science more broadly.[10] A social science that focused exclusively on the role of structures and conditions that were reflected in human practice, whether their practitioners were aware of them or not, had to be incomplete. While conditions were certainly causes, so too were the reasons people drew on when doing things. In other words, the investigation of meanings was part and parcel of putting human agents at the center of understanding.

And second, it was argued, the spatial–quantitative revolution and the major critical reaction to it up until that time, Marxist geography or historical geographical materialism, had ignored its significance, and this showed through in their deterministic views of the world: a spatial determinism in the case of the spatial–quantitative revolution and an economic one in the case of historical geographical materialism. This was not entirely true. Any approach to human geography drawing on historical materialism would sooner or later have to take beliefs and understandings into consideration, not least through the notion of ideology, and this had already been broached in Marxist geography in the pages of *Antipode*. Likewise, in the case of the spatial–quantitative work, behavioral geography had moved some part of the way in recognizing the role that people's perceptions of the world might play in how they related to it. One might certainly object to the positivistic way in which mental maps were investigated, but they were a recognition of the role of meaning.

Nevertheless, it was humanistic geography that would initially define the terms of the debate. The focus on meaning meant distinctive methodological protocols. Above all, it meant an emphasis on the hermeneutical, which refers to the imputation of meaning and is a necessary aspect of any scientific endeavor. In physical science it is a single hermeneutic: The investigator brings his or her framework of understanding about the nature of the objects being studied into a relation with those objects. To what extent do they behave, therefore, in a way consistent with the meanings one ascribes to them: the particular properties as they have been conceptualized? In the human sciences, a single hermeneutic will not suffice. Unlike the objects of physical science, human ones are also meaning-imputing and act in accordance with those meanings; as their meanings change, so their actions are likely to change. To talk about atoms in the same way would be absurd. In order to understand people's actions, one has to understand the meanings that material objects, practices, institutions, and texts have for them. This in turn means bringing their frameworks of understanding into a relation with that of the researcher. This is a circular movement. The investigator starts out with some assumptions about the frameworks of understanding held by the people being studied but then modifies those assumptions as inconsistencies emerge or new patterns take shape, and this will continue until there is some confidence that understanding has been achieved: that, for example, there seems to be some consistency between meanings on the one hand and institutional forms and practices on the other.

The emphasis, therefore, was on the interpretive. To the extent that contemporary cases were the focus, this was typically approached through ethnography: interaction with a specific community through participant observation, in-depth interviews, focus groups, the examination of particular events judged expressive of the sort of community it is. The process is an iterative one in accordance with the logic of the double hermeneutic outlined previously. A case or community is approached with a particular interpretive framework in mind. A variety of information is then collected, the bits placed in relation to one another to create some understanding of the context of meanings within which people are acting, followed by more observation according to that understanding, further adjustment, and so on.[11,12]

Few people think about humanistic geography any more.[13] Its contribution was overtaken by that of the "posts." From their standpoint, the analytic–interpretive distinction has to be given an entirely new meaning. In the case of humanistic geography, it mapped onto the structure–agency contrast. Given the antihumanistic stance of postmodernism and poststructuralism and their assumption of subjectification through discourse, this can no longer be the case. This does not preclude a critical interpretation of particular discourses, but it is always within the terms of some other discourse. Accordingly, the concept of truth that had previously prevailed in human geography, including that adhered to by spatial–quantitative geography, historical geographical materialism, and humanistic geography, has to be in question. Truth henceforth is relative to a particular discourse. In short, while there might be a material world, it cannot be known so as to arbitrate between one discourse and another. All one has is discourse.

How might one proceed in its investigation, therefore? According to Wylie, "Poststructuralist methods are above all *critical methods*" (2006, p. 298), and indeed that is where the overwhelming emphasis has lain, though constructively so. Accordingly, and when not swallowed whole, it has provided insights into human geographies that can be turned to effective use and provide "epistemic gain" in other sorts of research, including humanistic and Marxist geography. Two of its methods have proven particularly useful, though in practice they have tended to overlap. These are "deconstruction" and "discourse analysis."

Deconstruction

In deconstruction, the fundamental insight is that the meanings of objects are constituted by what they are not; by some "other." Behind the history of objects are discursive practices separating, excluding, and defining other objects as somehow lacking. The history of any object has, accordingly, to be construed in terms of its opposite: state–society, culture–economy, local–global, and so on. It is this that helps explain the deconstructionist fascination with the binaries that course through Western thinking.[14] In many of these, as in male–(female) or modern–(traditional), one term is assumed to be the original, the norm, as enjoying some sort of privileged status while the other is the abnormal, the derivative, the ineffective, and the lacking. In this way, identities get constructed. Stories are told to justify the distinctions and supported through material practice in a self-fulfilling sort of way.

Binaries of this sort abound in human geography: cosmopolitan–(provincial), space–(place), settler–(colonized), urban–(rural), London and the Southeast–(the rest of the United Kingdom) provide only a few of many examples. In some instances, they have been the object of sustained critical scrutiny. Much of Doreen Massey's *World City* is devoted to a critical examination of the last of these.

A contrast that, while useful from the standpoint of illustrating deconstruction, also has clear spatial referents is that between the indigenous and the nonindigenous. This is an extremely complex binary. It is one that has been of special interest to postcolonial studies as a way of distinguishing between a particular sort of nonindigene, the colonists and settlers on the one hand and those who were already there. Colonial encounters are the context within which the notion of difference congeals though at that time more in terms of a European–native binary to which all manner of other distinctions attached themselves, including ones of civilization–savagery, of race, and of the superiority of the "scientific" knowledge that the Europeans brought with them.

With independence, the indigenous–nonindigenous distinction acquired new meanings. Above all, it could now be used to legitimate privilege and precedence on the part of indigenous peoples: to justify the creation of land preserves exempt from land or timber concessions to Western corporations or to exclude them altogether. Here, it can join up with claims about the superiority of indigenous knowledge over that of those same—nonindigenous—corporations in materially appropriating a particular natural environment.

The old colonized–colonizer binary onto which the indigenous–nonindigenous distinction mapped is still there, but is now reinterpreted in terms of a rewriting of colonial history by independence movements: not as the bringing of light to the benighted but as an imposition that resulted in making them foreigners in their own land—an assault, in other words, on their status as indigenes. This, however, can be problematic, as I indicated in Chapter 4. In numerous instances in Africa, there are non-indigenous peoples who it would be difficult to assimilate to the category of "colonizer." It cannot be a simple African–white distinction. Numerous Indians came to the British colonies in southern and eastern Africa, partly as indentured workers, partly as traders. In what is now South Africa, slaves were introduced from the Dutch East Indies and from elsewhere in Africa, mixing with the Dutch settlers to create those who would become known as the Coloreds. In the highly racialized civil societies of the colonies, Indians and Coloreds came to form an intermediate stratum: less privileged than the white colonizers and settlers but more privileged than the supposedly indigenous Africans. This has created the context in contemporary South Africa for claims to rights that the intermediate strata, despite their history of subordination to colonial governments, cannot possibly have, and this in virtue of their nonindigenous status.

There are, therefore, observations that one can make that are disabling of the indigenous–nonindigenous binary, and largely due to the way it fails to map neatly onto some of the other meanings, like colonizer–colonized. Much more disabling, though, is the dubious nature of the binary on its own terms. The South African case is again useful here. Africans claim that they are the indigenous inhabitants, and this is something that has been legitimated by the preferential treatment provided by the African National Congress government. But many of them came originally from outside the current boundaries of the country as migrant workers for the gold mines. It is well known that there was a huge traffic between South Africa on the one hand and other parts of Southern Africa on the other. And many of them stayed.

More fundamentally, nobody is indigenous. Everybody is from somewhere else. Later waves of migration wash over the residues of earlier ones, creating all manner of hybridities: phenotypical through miscegenation, institutional, and in terms of knowledge. The idea of indigenous knowledge, like indigenous peoples, is a *reductio ad absurdum*. "Traditional" farming practice is a hybrid drawing on elements, like the iron plough and corn that were introduced from the outside. Institutions taken to be indigenous are typically far from it. So-called "customary law" is what legitimates the rule of the equally so-called "traditional authorities" in the rural areas of postcolonial sub-Saharan Africa. Yet as Mamdani (1996) has shown, customary law was a creation of the colonial authorities, drawing in a highly selective way, on the existing practices.

The valences attached to the binary, which side is valorized and which devalued, have therefore shifted over time. What has remained constant is the way in which it has always been drawn on in pursuit of what have been called (Seidman, 1992) practical–moral projects: projects that work in an exclusionary, subordinating manner. And while I have been concentrating on the colonial case, it is a binary that clearly has much wider resonance as in the claims of citizens versus immigrants, nationals versus cosmopolitans, or any "insider" as opposed to those "outsiders," typically identified as "trouble makers," who try to impose their views on the innocent. Yet again, just as the independence movements reinterpreted it to suit their purposes, so too it can be read in a different way: that is, one where the local–nonlocal maps onto such binaries as incurious–curious, unenterprising–enterprising, timid–adventurous, unimaginative–resourceful. The very fact that the negative pole of the binaries comes first here underlines just how unsettling this tactic is.

Discourse Analysis

Discourse analysis is a closely related approach. It involves the critical examination of texts or narratives of all sorts, including maps, geographical writings, landscape interpretations, and media reports for their underlying grammars, their hidden assumptions, their silences, and the particular classed, raced, gendered, or whatever, viewpoint that they represent. Whether they set out to or not, texts define certain practices, institutions, and identities as normal, even natural, while others seem to be unusual and abnormal. Through the implicit rules that govern their structure, discourses determine what questions can be asked and the legitimate range of problems to be investigated.

As Derek Gregory has pointed out (Johnston, Gregory, Pratt, and

Watts, 2000, p. 180), the analysis of discourse opened up new possibilities in the understanding of geographic thought, though one could make the same point about deconstruction. The writings of any discernible school—the Sauerians or even, of course, of postcolonial geography or particular approaches like the new geopolitics as well as its classical precursor—can be fruitfully examined in these ways. In Chapter 9 I examine the spatial–quantitative revolution from this viewpoint. Likewise, before the development of feminist geography in the mid-1970s, writings in human geography were, in retrospect, quite extraordinarily masculinist. The presence of women might be acknowledged in population geography as in sex ratios, but otherwise hardly at all. Economic geography did not differentiate between female and male wage employment, commuters were all commuters, and writings about the living place to which women in the developed world were still largely confined up to the middle of the last century were remarkably thin.

These are all cases of what have been called imaginary geographies. But clearly geographers are not the only ones who have them, and as a result of the way that they subjectify, they are a crucial tool in interpreting, and critically so, contemporary human geographies. The public world of local and regional development in the United States, the world in particular of politicians, planners, and so-called "local economic development professionals" has its own language, its own classifications of that space and those spatial processes into which it hopes to intervene. And political as it necessarily is, since this is a politically highly fraught area, that imaginary geography has its own very distinct silences.

To illustrate further, in the United States local and regional growth is an obsession, vigorously promoted by local and state government in alliance with a variety of private interests: in particular, utilities, developers, and property companies and sometimes banks and personified in the form of the local economic development professionals. The discourse in which they trade includes a number of key, mutually reinforcing ideas. As such, they provide what has been called "a regime of truth" that can inform the promotion of, and if necessary the struggle for, particular development initiatives. Its elements include:

- *"Territory."* Nobody speaks explicitly of territory, but they do talk about "our great state of Ohio," "our city," even "our neighborhood" and sometimes "our region," implying that there is some shared interest among those living there with respect to its relations with people, institutions, and agents elsewhere.
- *The interest that is shared is in "development."* The word "development" is crucial, as in the "development" of "our town/state." On the one hand, who could oppose "development"? It speaks of "improvement," of the betterment of individual lives. It is, moreover, the "development"

of an area or place that is at stake. Everyone stands to gain. To use the standard metaphor, it promises to "raise all boats." A moment's reflection will suggest just how odd this is. In what sense can an area develop? Areas are, rather, in this instance at least, containers for different agents—workers, local government agencies, firms—each of which can defensibly be defined as "developing" through the acquisition of increased capacities and wealth, either individually or through various forms of association with one another. A place might also develop in the sense of realizing a particular identity in which all can share and the spokespersons for it are indeed quick to talk about its implications for "becoming a major league city" or "putting our town on the map." But again, it is an identity for individuals and not the areas in which they live.

• *"Places" "compete" for "development."* Since all places want the same things—government infrastructural projects, the inward investment of firms in the form of new factories and offices—and it is assumed that these are in short supply, then "places" must "compete" with each other; territorial competition becomes a *leit motif* of the local and regional development discourse.

• *The way that they typically compete with one another is through the creation of a suitable "business climate."* This means an ensemble of low taxes, minimal bureaucracy, pliable workers who refrain from striking, that investors in factories, logistics centers, back offices, even hotels, will find attractive. And since we all want "development," it is incumbent on all of "us" living here to make sure that these inducements are in place. Other places, it will be urged, have experienced decline, and a major reason for that is that they allowed labor unions to get too powerful—a warning for "us" all.

• *What can upset this, even when all steps have been taken to ensure an attractive "business climate," is the failure of the state to ensure "a level playing field."* Through legislation, states can, whether they are aware or not, advantage some places over others. Through the tax system and subsidies from the state to constituent cities and regions, some can get more than their "fair" share: more infrastructure, more state employment to the benefit of particular city housing markets, and so on. Statewide legislation that is indifferent to local variations, like a minimum wage law, can have the same effect. It is as if, and it will be sometimes claimed to be the case, some cities or regions "exploit" others.

This is a set of claims that is endlessly communicated through the public statements of politicians and through the media. Sometimes in newspapers it is through the synthetic, more abstract statements of the editorial and op-ed pages. But the reports on individual rezonings, city council hearings on infrastructural initiatives, and pending state legislation are replete with references to implications for "development" or "business climate" and "creating a city we can be proud of." In the United States, at least, even the most unlikely issues are forced through the local and regional development prism. Legislation to limit the confinement of farm animals and poultry can become a threat to jobs in "our" state, all accompanied by statements as to the taxes paid by farmers and so benefiting "all of us." The same goes for air pollution legislation and what it means for coal mining.

A product of this discourse that is of particular interest and that in turn produces further discourse is the creation of a set of imaginary geographies that then circulate through the media. These may include maps of relative performance showing how "our" state is doing relative to others—always relative to the "competitors"—in terms of population growth—a typical sort of news commentary when census or interim-census counts are announced. There are also geographies, perhaps mapped and perhaps not, which deal in a simplified geography of oppositions: Cold Belt–Sun Belt–the "other" Ohio,[15] central city–suburbs but also inner suburbs–outer suburbs.

Central to the creation of these imaginative geographies are the various agencies representing the development interests of particular states, but also the consultants who thrive on the intense desire of local governments and growth coalitions to "develop." In some cases, states or the growth interests located in them have come together in order to push agendas that are more regional in character, sponsoring "research" organizations to provide them with the necessary lobbying ammunition: organizations like the Northeast-Midwest Institute, which is sponsored by the Northeast-Midwest Congressional Coalition, a bipartisan group that lobbies for the so-called Cold Belt states; or something like the Center for the New West, which has a strong conservative edge.[16] The consultants can likewise be extraordinarily influential. The person who springs readily to mind here is Richard Florida and his thesis of "the creative class" as a tool in local development. But there are numerous others, including Joel Kotkin, a strong advocate of traditional suburban development and Myron Orfield, who has helped embed the inner–outer suburb opposition in the consciousness of development interests in the older suburbs of the country.

The imaginative geographies so created are subject to high levels of oversimplification, and necessarily so, since otherwise the stories that they tell would be less compelling. The easy contrasts indulged in mask major issues of geographic scale: To put it in analytic terms, to what extent, for example, can it be said that the variability in developmental outcomes across counties in the United States is largely "accounted for" by the Cold Belt–Sun Belt contrast? To what extent is it a matter of individual states or even of broader regions not easily accommodated to the Cold Belt–Sun Belt distinction? Likewise, since the object is to create some sense of territorial variation, the various ways in which the different territories are interlinked, mutually dependent for their fortunes, tends to get glossed over. So while indeed federal procurement orders might benefit some aerospace contractor in Los Angeles (i.e., a Sun Belt location), since that is where final assembly will occur, what gets assembled is produced by numerous subcontractors scattered across the United States, including in Cold Belt locations.

It is, in short, a discourse for some people, working to realize their particular agenda, and not everyone: again, what Seidman (1992) has called a "practical–moral project." It is the local and regional development lobby that has been central to its elaboration and dissemination, and the media have bought into it, particularly where, as in the case of many of the local media empires scattered across the United States, they happen themselves to have strong interests in the growth of local or regional markets. Like all discourses, it works through its selectivity. Other stories, equally if not more convincing, do not get told; in fact, to tell them would be to subvert its purpose of subjectifying people in certain ways, softening them up to act in ways, like voting in referenda, that will further the local and regional development purpose.

It is not difficult to develop an alternative discourse; it does not have to be that radical, and it occasionally surfaces as a minority voice, a small minority voice, in the context of particular development issues. This is one that emphasizes the redistributional effects of developer and growth coalition activity and self-consciously positions itself as on behalf of "the people": the impacts on local property values and environmental amenity; calling into question the claims of the jobs that will be added to the local economy when Walmart comes to town but failing to mention the jobs that will be lost; and arguing against taxpayer-funded infrastructural investments that will benefit a few property owners and developers. In other words, discourses are fashioned with respect to one another. They aim to create not just alternative forms of identity but dominant ones that will edge the rivals aside. The discourse of local and regional development that tends to dominate in the United States is, therefore, inconceivable outside the shadowy presence of its "other" and where it has to it will respond to that "other" in its own terms, identifying the "trouble makers" as "outsiders" whose effect will be to damage the local "business climate" and deprive "us" all of the benefits of "development."

CATEGORIAL–DIALECTICAL

The sense that the world is constantly changing, the ephemerality of things, can be readily assumed. It is all the more surprising, therefore, that it has taken so long in the social sciences, including human geography, for this to be registered in terms of method. As sociologist Martin Albrow pointed out in 1974, the assumption in his field was one of order and stability in the world; change was simply a reworking, a reordering, perhaps some changes in relative magnitudes in what already existed. There has been some change in viewpoint since then, but it remains the fact that a self-conscious hewing to what Albrow advocated instead—what he called a "dialectical paradigm"—has remained a minority interest. In geography a particular form of the dialectic, the materialist dialectic of Marx, was evident 40 years ago in Harvey's (1973) *Social Justice and the City*. However, it is only relatively recently that there have been attempts to set out exactly what dialectics entails, notably by Harvey (1996, Chapter 2) himself. I want to start, therefore, with some of the essential aspects of a dialectical approach, though the way in which they form a unity is crucial.

First, there is the prioritization of flows over things. It is to adopt a particular understanding of the idea of process as the dominant moment in the world: how, in Harvey's words, things—factories, cities, people—should be seen not as fundamentally thing-like, as stable, enduring, as the essential elements in an understanding of the world, but as, as he put it, as "permanences": only *relatively* enduring and sustained by flows and relations, therefore—sustained, in other words, by processes like the various natural cycles, hydrological, nitrogen, and so forth; by the circulation of knowledge and people and means of subsistence; and not least by that of capital, which determines whether factories get re-equipped, repaired, demolished, or left to decay. Things are merely moments in processes. They are more or less temporary forms of storage of water, nitrogen, minerals, capital, and knowledge, capable of mobilizing sources of energy and turning them to various purposes. Nothing is exempt from this view. Institutions like states can and should be imagined as "permanences" as officials come and go and revenues in various forms—taxes of different sorts or appropriations in kind—flow through the state and allow its reproduction. In their turn, state officials then make production possible through the provision of various forms of infrastructure, physical and social. Accordingly, "permanences" are defined relationally: through their relations to yet other "permanences," whether it is capital and wage labor, metropoles and colonies, or the urban and the rural.

This might convey the idea of stability—in other words, a constant recycling of water, nutrients, capital, people, through various "permanences" so that they simply get reproduced or replaced from one point in time to the next. Dialectics, however, is, as Albrow was so keen to emphasize, about change, and change of a particular sort: qualitative change. Change can indeed be quantitative: more people, bigger cities, the changing balance of different species, the movement of people from rural areas to cities, increased trade, a deepened division of labor. More fundamentally, from the standpoint of new causal forces, it is also

qualitative: new sorts of "permanence," in the form of new social institutions, technologies, understandings, and new forms of spatial ordering. Nothing is entirely new; it always builds on what is already available,
what "permanences" already exist. But through their combination, new
forms of capacity, latent with implications for further change, are generated. There are emergent powers, in other words. Likewise, this is not
to radically set the qualitative against the quantitative. As quantitative
thresholds are achieved, "permanences" can change in their properties:
Migrants create "homes away from home," cities acquire new functions
hitherto unimagined. Qualitative change, on the other hand, can release
new forms of quantitative change, as in those unleashed by the advent of
capitalist production relations.

These changes do not just happen. Mediating between flows and qualitative change is contradiction. Through their relations with one another,
"permanences" encounter obstacles to their reproduction.[17] Negotiating
these barriers then results in novelty. Examples are easy to find in human
geography. Increasing populations have been viewed as conditioning
the development of agricultural practices: changes in agricultural technology, therefore (Grigg, 1979). With increasing geographic extent, the
mononuclear city gives way to a multinuclear form. As the world becomes
qualitatively richer, more differentiated, as new sorts of "permanence"
appear alongside older ones, so new tensions emerge latent with further
transformation. New modes of production enter into conflict with older
ones. The unity of the family disintegrates with the formation of new life
cycle stages antagonistic with respect to one another and with implications for the residential structuring of the city.

Table 6.2 identifies some of the leading questions to be asked when
drawing on a dialectical approach to the world. This understanding of the
dialectic can be usefully contrasted with others, which stray from its relational core in the direction of more reified understandings of relations:
one in which things precede relations, relations through which things are
then changed in a reciprocal sort of way. Soja's early ideas about what he
called "the socio-spatial dialectic" fell into this trap. His approach was
ambiguous, flirting with more truly relational views—space "represents
. . . a dialectically defined component of the general relations of production, *relations which are simultaneously social and spatial*" (1989, p. 78;
my emphasis)—but then lapsing into a world of separate parts that interact and form one another—"social and spatial relations are dialectically
inter-reactive, interdependent; . . . social relations of production are both
space-forming and space-contingent" (1989, p. 81). It is, however, a snare
in which people continue to find themselves entrapped. In the entry on
Scale in the latest edition of *The Dictionary of Human Geography* (Gregory
et al., 2009, p. 665), we learn that "this recursive relationship between
social processes producing scales and scales affecting the operation of

**TABLE 6.2. The Dialectic as an Approach
to Asking Questions of the World**

- How are things changing, being transformed, decaying? What new functions are they performing in wider sets of relations, and what other functions are they no longer satisfying? How are their meanings being transformed?

- What does this have to do with changing relations and flows?

- What are the processes in this instance?

- What are the ensembles of relations, the wholes, of which something is a part?

- How are changing relations and flows conditioned by contradictions in the world and what are these contradictions?

- How does this affect the relation of parts and wholes? How are wholes constituting parts and vice versa?

social processes is one aspect of the socio-spatial dialectic." Quite how scales might be able to "affect" goes unremarked.

This shows perhaps just how difficult it can be to liberate thought from what Albrow defined as the categorical paradigm: as grasping objects through reified categories so that they are divorced from any wider process that might help account for their very existence, let alone preempting the possibility of dissolution, transformation, or transcendence (Table 6.3). Things do not change each other and then get changed in turn. Rather, they internalize relations with other things from the beginning. Without those relations they would not exist.

Just how different a dialectical understanding of the world is can be obtained by a consideration of some central contrasts (Table 6.4). An awareness of change is not a monopoly of dialectical method, but the sort of change it emphasizes is. Many depictions of change are of a purely quantitative sort: the growth of markets and of the state, improved hygiene, the decline of the peasantry, imperial expansion, and so on. Capitalism is the expansion of the market rather than a qualitatively different

TABLE 6.3. Categorial Frameworks and Leading Questions

- What are the things (variables) under study?

- How can things be classified? What are the appropriate pigeon holes?

- How are they patterned with respect to one another? What are the forces generating equilibrium among them?

- How are those patterns and equilibria affected by the inclusion of new variables?

**TABLE 6.4. Categorial and Dialectical Approaches:
Some Essential Contrasts**

	Categorical	Dialectical
Change	Quantitative	Qualitative
Basic modality of existence	Being	Becoming
Ontological priority	Things	Flows and relations: things as processes or relations; "permanences"
Origins of change	Contingent	Necessary
Meanings	Stable	Shifting
Situatedness	Unaware	Aware
Mode of abstraction	Empiricist	Systematic or historical
Time–space applicability of concepts	As universal as possible	Specific to geohistorical contexts, but in dialectical relation to the universal

set of relations, hitherto unencountered, between producers and what they work on and with. Rather, things—the relations between things—are unchanging; they do not become something different but simply "are." Things are elemental, the basic stuff of the world: durable, self-sufficient, existing in and of themselves—"being," in other words. They then enter into relations in accord with their enduring and essential nature: distance-minimizing movements, since people are rational; trade, because they have an inbuilt propensity to truck, barter, and exchange or, if it is countries, because they differ in their resource endowments. The alternative is to prioritize the relations that are presupposed by things: to understand human nature as something constructed in virtue of a particular set of social relations, landscapes as differentiated flows of money, creating pleasing prospects here but shabby working class settlements elsewhere.

For a categorial view of the world, change is something that is possible but not necessary. It depends on conditions that cannot be anticipated: the revolution in telecommunications and globalization; the rise of the newly industrializing countries and the creation of "rustbelts"; the invention of the automobile and low-density suburbanization. The alternative is more nuanced: to recognize the unanticipated nature of the specific changes that occurred but also the fact of change itself as omnipresent and as something to be explained. Rather, change is necessary, driven by the contradictory nature of the relations into which things necessarily enter; "necessarily," since they are not self-sufficient but depend on those relations for what they are. As change occurs, so meanings change. The world assumes new contours, which confound representation in the old ways. New terms, or terms that would have been new at some time, make their appearance: for example, Sun Belt, high tech, Cadillac desert, economic development. And old ones like civilization and masculinity get reworked. What might appear to be stable elements of the world are continually entering into new relations, being reworked, given new meaning, further differentiated in their meanings.[18]

To engage in dialectical thinking is to recognize the situatedness of all ideas and practices, since it is to recognize how the thinking person is him- or herself constituted by a bundle of ever-changing relations with others through social institutions, discourses, divisions of labor, and the like. If it is to be understood, everything should be grounded in terms of the necessary and changing conditions of its becoming. For a more categorial view of the world, the issue of situatedness does not arise for there is nothing to situate. Things are what they are or, as the financial press is inclined to aver, "It is what it is." This transfers over into very different understandings of the process of abstraction (Cox, 2013a).

Finally, there are the sorts of concepts that are favored. The underlying assumption of categorial approaches is that the world is stable and orderly. What exists has always existed in some form or other. Given this assumption, it is not unreasonable to develop and work with concepts that have a seemingly universal applicability, as in those of human nature referred to previously.[19] The common objection to this is that, as a result, the concepts lack specificity and are, therefore, only minimally explanatory. There has always been a division of labor, but it matters whether it is a division within the family or mediated by commodity exchange; whether it refers to the division between the production of one final item of consumption and another or between the people who are part of the process dedicated to that particular line of production. This is reasonable enough, but dialectics facilitates greater critical penetration into the matter.

The tendency in dialectical thinking is not to reject the universal but to see it in terms of its relation with the particular; how the universal nature of the universal is only revealed by the differentiation of things as they develop. It is that complex expression that then allows the nature of the universal to be grasped: "As a rule, the most general abstractions arise only in the midst of the richest possible development, where one thing appears as common to many, to all. Then it ceases to be thinkable in a particular form at all" (Marx, 1857–1858/1973, p. 104). In short, the universal is a construction which only makes its appearance under definite socio-historical conditions. The concept of space is a good example, dependent as it is on the sorts of abstractions from place that the rise of capitalist commodity exchange facilitated and encouraged.

What *appears* to be universal at any one time, therefore, bears the mark of the concretely geohistorical. Through a materialist form of the dialectic, Marx saw the understanding of the universal as something that developed over time. So in the case of money: "This very simple category . . . makes a historic appearance *in its full intensity* only in the most developed conditions of society" (p. 103; my emphasis); and "although the simpler category may have existed historically before the more concrete, it can achieve its full (intensive and extensive) development precisely in a combined form of society" (p. 103). Universals are understood relationally and those relations change over time: "Money may exist and did exist historically, before capital existed, before banks existed, before wage labor existed, etc. Thus in this respect it may be said that the simpler category [in this case, money—KRC] can express the dominant relations of *a less developed whole*" (p. 102; my emphasis).

One can assume that our understanding of space as a universal has an analogous relational history. The idea of relative location must be, as Marx states with reference to labor in general, "immeasurably old" and certainly, judging from the existence of maps both in precapitalist Europe and pre-Columbian societies, antedating the emergence of a form of society in which self-sufficiency gave way to a chronic dependence on exchange. The idea of locations as substitutable for one another is in all likelihood much more recent, perhaps no older than the full blossoming of property capital as a distinctive part of capital's division of labor. What space means as a universal category has been enriched over time as relative, and then relational concepts of space have taken their place alongside more absolute conceptions, though quite what that history would be in its details remains to be written.

The spatial–quantitative work of the 1960s was categorial in its approach to a fault. Striking in hindsight was the search for concepts of maximum time–space scope: concepts like settlement dispersion, place utility, search space, formal and nodal regions, or key relations like Tobler's law and distance decay. The geometrical approach outlined by Peter Haggett in his *Locational Analysis in Human Geography* carried this to an extreme, starting out with a highly abstract set of spatial concepts that might seemingly be applied to all societies at all times: movements, nodes, hierarchies, networks, and surfaces. This is a vision equally apparent in Bunge's *Theoretical Geography*.

One is also struck by the emphasis on the orderliness and stability of things. Space was seen as "patterned," "organized," and "predictable," if only the principles of that ordering could be uncovered, which, of course, they could be and often would be through the quantitative isolation of different components of variation. To the extent that change entered the picture, it was a matter of changing patterns: changing gravity exponents, shifting connectivities, variations over time in degrees of localization or segregation, or the difference that changing transport technologies might make. Qualitative change was seemingly foreign to this way of thinking.

The actual nature of things was subordinated conceptually to their quality as varying. Geographies were accordingly to be explained by geographies, the variation of one thing by the variation of another, differences of degree by differences of degree: urban size by interurban distances, migration between places by the (geographically varying) terms of the gravity model. The more relational mode of thinking associated with dialectics was a long way in the future.

This is not to downplay the utility of this approach. It has indeed embraced concepts that have been usefully applied across a wide swath of time–space. Archaeology, in particular, benefited a great deal (Hodder and Orton, 1976; see also Tobler and Wineburg, 1971). Central place theory has been a very powerful tool in the understanding of settlement patterns wherever there is exchange. The approach also captures something that is fundamental to the way in which modern industrial societies are organized geographically: the variable connectivity of cities to airline networks, the quite extraordinary power of the gravity model with reference to migration and trade, testifying to the continuing significance of concepts of space as relative. What it failed to do, though, and what a more dialectical approach might have offered, was to situate itself with respect to what else was going on in the world: folding space as relative, therefore, into space as relational, seeing the idea of space as relative in relation to the demands of city and regional planning, the technocratic mood so dominant in the 1950s, and the dilemmas of a discipline fighting for survival.

PLURALIZING–TOTALIZING

The categorial–dialectical distinction overlaps with another significant methodological contrast: that between pluralizing and totalizing approaches in the human sciences. Once again, it departs from another slant on the constitution of the world. To talk about a pluralizing approach is to assume a world composed of a set of independent forces that interact with one another, creating different conditions within which people, as one more distinct force, interact. Little is said about change, whether it is necessary or not and its direction—at its extreme a pluralistic view postulates a future that is entirely open. It is an approach that has been very common in human geography.

In spatial–quantitative work, a common technique has always been multiple regression analysis. This attempts to account statistically for the variation in some "dependent" variable by reference to what are called "explanatory" or "independent" variables. The word "independent" is crucial. The explanatory variables are assumed to be "independent" of one another. If there are strong correlations between them, then this creates instability in the values of the different regression coefficients[20]; there are, in consequence, techniques designed to check for this problem of "multicollinearity" and to assure the investigator that they are indeed truly "independent" of one another.

The different systematic fields and their separate development as economic, cultural, political, and social geography also bear witness to this pluralized understanding, and it has been a continuing theme in much of the literature. One might recall here the distinction made by critical realism between the necessary and the contingent or the heterogeneity of the world at the center of the postmodern approach. Subsequent to the Marxist geography of the 1970s, its totalizing understanding of the world allowed pluralist approaches to position themselves in new ways. This has been true of postmodernism and poststructuralism and those, like Massey, who have been strongly influenced by them. It has also been common among feminist geographers to reject the claim that patriarchy is one of capital's production relations in favor of assuming its independence.[21]

The lineage of pluralizing approaches is a complex one. The multivariate analysis of the quantitative geographers is clearly very different from the emphasis on different structures of social relations central to critical realism. Postmodernism and poststructuralism are different again. Multivariate analysis offered a highly "thing"-ified view of the world: a categorial view to a fault. The relationality of critical realism went some way toward undoing that, but it was a relationality that tended still to break the world up into fragments in the form of diverse structures.[22] Deconstruction of dualistic thinking has also seemed to offer promise, but it is a promise that remains unfulfilled given its refusal to acknowledge that knowledge might have foundations outside of itself. All views that ascribe some unity to the world are seen as partial and irremediably political.

An alternative is to see the social world, at least, and in its relation to nature, as a totality. In human geography, this is entailed by a particular version of the dialectic: that of Marx. This is a view that is widely misunderstood as deterministic and to be more specific, economically deterministic. One can certainly acknowledge that it is a materialist understanding that locates production, albeit *social* production, at the center of its worldview. Nature and other aspects of social life then become conditions of production or production relations. This is the basis of historical materialism's strong holistic tendencies.

This is far from saying that it is deterministic, though. Concrete futures cannot be read off from the present. The social world, including relations to nature, is transformed through the sequential posing and suspension of contradiction. Contradiction might seem to be quite remote from the sphere of social production—a contradiction between political forms and imaginaries, for example. Their resolution, however, must always be in terms of the limits, discursively mediated for sure, of social production: Under capital, this means the need to grow, to accumulate, something that is given *not* discursively but by material relations between people and people on the one hand and nature on the other.

A particular flashpoint of these debates has been in feminist geography, where the crucial question has been the degree to which gender or patriarchy constitutes what Massey described in a 1991 polemic targeting, among other writings, Harvey's *Condition of Postmodernity*, as a separate "dimension of power": separate, that is, from capital. The most clearly argued justification for this claim was that of Foord and Gregson (1986), who did so in quite explicitly critical realist terms. Gender appears as a separate structure of relations of a transhistoric kind, which, when combined with capitalism, yields patriarchy. Hence:

> To analyze gender relations as a conceptual category . . . does not require us to analyze patriarchy. Nevertheless, the inverse does not hold: just as we need the general category mode of production to analyze capitalism, so we need the general category gender relations to analyze patriarchy. So gender relations are necessary to an analysis of patriarchy, but patriarchy is contingent to the analysis of gender relations. (Foord and Gregson, 1986, p. 200).

In response to this, we should first note that there can be no gender relations without production. The married couple is one sort of gender relation, but it is hard to imagine it being reproduced without some sort of access to the means of subsistence. Of course, one could modify one's position and say, as do Sayer and Walker (1992), that gender relations are one form of the division of labor. But there again, the division of labor requires some set of material exchanges to get it going and to sustain it: The division of labor in this sense presupposes labor, and that requires an access to the means of production, which, as history affirms, cannot be taken for granted.

One might reply that this is not exactly what is meant when referring to a contingent relation between gender relations and production or mode of production. Rather, it is that a particular mode of production does not correspond to any particular form of gender relations, and this is certainly true. But that does not mean that there is an infinite number of forms that they can assume given a particular set of production relations. Rather, they have to be such as to facilitate, and not impede, the particular logics of production entailed, whether it is slavery, feudalism, capitalism, or some hybrid set of relations that we are talking about. This is what Harvey was getting at in his reply to Massey in 1992 when he argued that "criticism and analysis can(not) proceed as if contemporary women . . . are situated outside of the shaping powers of capitalist society itself." Again, it is clear from the history of gender relations that as the needs of production and its necessary conditions have changed, so too have they.

Not all forms of totalizing argument have been as controversial as this. The emphasis on relations between parts hitherto viewed as separate can be a highly productive tactic, which raises few eyebrows. One thinks here of the way Michael Storper and Richard Walker brought together studies of industrial location and regional development in their (1989) book *The Capitalist Imperative* or Brian Page's (1996) work on agriculture and industrial geography.[23]

CONCLUDING COMMENTS

In conclusion, we should note how particular methods in the narrow, technical sense have become associated with particular epistemologies and, therefore, with particular understandings of social life. The obvious case is quantitative methods and the positivism that was the spatial–quantitative revolution: a positivism that, in turn, jibed nicely with the "thing"-ified notions of social process expressed in location theory and in particular that of individuals governed by markets, sliding down price gradients but in turn influencing those markets through their preferences for, for example, particular transport cost/rent budgets. This is far from the only case. Discourse analysis and deconstruction obviously have a very close connection with poststructuralism and postcolonialism.

But in neither of these instances is the connection a necessary one, which is one of the reasons that I decided to treat questions of method in a separate chapter. As I emphasized earlier, and as Sayer has underlined, quantitative methods can find application in *any* study seeking to isolate cases for more intensive study or seeking to establish relations as the object of investigation. The recent tendency of the legatees of the spatial–quantitative work to move away from the law-seeking approaches so closely associated with its origins in the direction of a greater appreciation of context is another case in point. Some of the work of Fotheringham (1981a)[24] is exemplary here. Likewise, there is no reason discourse analysis should not find a place in work framed by historical geographical materialism. The example I gave of local economic development discourse does not necessarily entail a poststructuralist understanding of how social life proceeds. It is, rather, entirely in accord with the fetishized notions of space that we, in virtue of our embeddedness in a capitalist society, all share and that, therefore, makes us so vulnerable to the blandishments of growth coalitions.

A final and perhaps less obvious instance concerns the structures of social relations that Sayer puts at the center of the causal structures privileged in his intensive forms of method. In critical realism, the identification of the structures proceeds through the separation of the necessary from the contingent. A consequence of that in terms of ideas of social process is pluralizing—so many independent social structures like capital, the division of labor, gender, and so forth—putting critical realism in an uncomfortable relation with the categorial approaches discussed previously. Yet the creation of new structures of social relations so as to facilitate certain desired effects is in no way foreign to capitalist practice or to working-class practice for that matter. Rather, it is a characteristic way in which contradictions are confronted: the reworking of old structures, the creation of new ones, or movement to take advantage of their presence elsewhere. There is, in other words, no reason in principle why the idea of structures of relations, albeit viewed as capitalist forms of structure, should not be allied to the dialectical, totalizing vision of historical geographical materialism.

In short, it is not difficult to pick apart methods along the different dimensions discussed previously, and connect them up fruitfully with research in theoretical traditions where their presence might at first blush seem inappropriate, and much still remains to be done in this regard. There are clearly limits to this. No Marxist geographer is going to use multiple regression with a view to establishing the independent causal significance of different variables, since it is counter to a totalizing understanding of the world. But drawing on measures of the localization of different industries with a view to investigating geographically uneven development would seem entirely acceptable.

NOTES

1. Compare Sayer (1979, p. 19): "In geography's 'philistine' period—that is at the peak of the so-called 'quantitative revolution'—geographers happily turned their backs on traditional geography's amateurish reflections on the philosophy of the subject and the relations between 'man' and Nature and contented themselves with a zealous but narrow concern with scientific methodology." One should, however, take exception to designating the spatial–quantitative revolution as "geography's 'philistine' period": while human geography might yet have discovered method and theory as objects of contention, without the spatial–quantitative revolution it would have been delayed.

2. This has tended to change more recently. Qualitative methods are now taught, there is the occasional article on topics like ethnography (Herbert, 2000) and there is a *Handbook of Qualitative Methods in Geography*.

3. Taaffe (1974, p. 8) provides a nice summation from inside the spatial–quantitative revolution: "On the positive side, the spatial view seems to have been more productive of cumulative generalizations than its predecessors. As the studies progressed, the emerging generalizations showed a certain amount of consistency and cumulation. The findings of one study were starting to be used in the next study. In the late fifties much attention was given to classical

theories and ideas related to spatial organization. These earlier statements were subjected to considerably more rigorous empirical tests than had previously been the case, and were reformulated for further testing."

4. For example, the event at the first point in time could have been initiated in the knowledge that it would produce the second event. So was it the first event that *caused* the second one, or was it the understanding that doing X at the first point in time would result in Y at the second point in time that was the causal condition?

5. For an example of how this works in practice, see Cox and Wood (1997).

6. On the question of the researcher's positionality, see England's (1994) excellent discussion.

7. Cybriwsky (1978) provides a nice illustration of this approach.

8. Compare David Ley (1988, p. 131): "The qualitative-quantitative dualism separates unnecessarily hermetic categories which, like others in human geography, should profitably be integrated together."

9. Richard Morrill's work should also be noted. He extended Hägerstrand's method to a number of different substantive areas, including the expansion of the black ghetto (1965b), peripheral development around cities (1965a), and the development of settlement patterns (1963). None of this was what one might call "descriptive generalization" research, even while such generalizations might be drawn on in developing some idea of appropriate process.

10. This was so even while humanistic geography could have shown more interest in interpretive social science than it did. See Sayer (1989a).

11. Steve Herbert's (1997) study of the territoriality of the Los Angeles police department is a fine example of participant observation research. Roman Cybriwsky's (1978) study of a multiethnic, working-class Philadelphia neighborhood provides another excellent example of an ethnographic approach. As he states: "It is based on the participant observation method of field research, and its data include the conversations, incidents and events which characterize the neighborhood and its residents" (p. 17). This was a lengthy field experience; altogether Cybriwsky spent 4 years in the neighborhood at four different locations. Just what the neighborhood means to residents is a major focus and the range of observations drawn on in documenting the high level of neighborhood cohesion impressive: the sort of evidence that could, indeed, only be gathered during a lengthy period of time.

12. For a good discussion of the process as it applies to historical data, see R. C. Harris (1978). Note, however, that the "historical mind" that he recommends is not just to be recommended to humanistic geographers. It has long been practiced by historical materialists, as in the signal contributions of E. P. Thompson (1967, 1971).

13. Note, though, that some of its concerns have recently been revived under the heading of emotional geography.

14. For the deconstructions by geographers of a number of binaries common in the human geography literature see Cloke and Johnston (2005).

15. This received some publicity some years back: the "other Ohio" comprised those parts of the state that were disadvantaged in the location of state employment. In particular, it was a call for the redistribution of state employment, particularly away from the state capital in Columbus.

16. From the Center's webpage: "The mission of the Center for the New West is

to foster balanced growth and economic development in the Western United States and to advance the principles of the Founders, including personal freedom, limited government and free enterprise in America's New Economy. Founded in 1989, the Center seeks to improve awareness of Western issues and to provide a platform for leaders of business, government, education and civic institutions to address regional, national and global problems and to develop practical solutions for them." The appropriation of "the principles of the Founders" is especially interesting.

17. A reproduction that can require quantitative change, as in the expansion of a firm.

18. To refer to a political party as "right wing" or "left wing" has been utterly transformed in comparison with half a century ago. The same applies to working class as a lived experience, as the blue-collar component of working-class employment has declined along with "traditional" occupations and occupational communities like mining villages, steel towns, and fishing towns.

19. Compare Albrow (1974, p. 190): "Whether it be structuralism or recent systems theory, with its language of control, feedback, input and output, the result is a framework of concepts of immense generality and scope and empty of specification."

20. If there were two independent variables, $X1$ and $X2$ each would have a coefficient resulting from solving the multiple regression equation. The coefficient would indicate how much change in the respective variable would result in a unit change in the dependent variable. In the presence of multicollinearity, the values of these coefficients could be highly unstable, so that a different sample could produce very different results.

21. See, for example, Foord and Gregson (1986) but also McDowell's retort (1986).

22. See Gough and Eisenschnitz (1997) on the attempt of Sayer and Walker to develop the case for the division of labor as an independent structure of social relations in their (1992) book *The New Social Economy*.

23. Another instance of the search for unity in the seemingly disparate was Harvey's (1978) demonstration of the unity between struggles in the workplace and living place: how capitalism had separated the two in physical terms but how the accumulation process was the necessary condition for both of them. Accordingly, the living place had become a field for accumulation on the part of developers and financial agencies, bringing about popular resistance to the various environmental impacts of the former and the redlining policies of the latter.

24. See also his review paper (1981b) on this topic.

CHAPTER 7

Human Geography
and How and Why Things Happen

The questions of why and how things happen, the question of causation, are central to any science, human geography included. Threading their way through the history of the field have been a number of different attempts to come to terms with it. Much of this has been carried on through a debate about the relative roles of human agency and structure, the latter variously understood as ecological, social, or socioecological. This seemed to have been resolved by the early 1980s. Hitherto structure and agency had been seen as independent forces. Henceforth they would be conceptualized in relational terms: Agents could only be agents in and through particular structures of social or socioecological relations; structures, therefore, had to preexist any particular agent—yet, on the other hand, they only existed in virtue of human activities.

Quite what effect this had on the research of human geographers is unclear. On the one hand, the relational formulations of the people proposing them, like Bhaskar and Giddens, were at such a high level of abstraction as to offer little guidance. On the other, there is the question of the degree to which they were simply developing a solution that was "in the air" anyway, at least in human geography. And after early worries about how to apply Giddens's structuration model, recognition of the relational character of agency and structure became common in human geography research and without any self-conscious reference to the debates of the early 1980s. If this particular resolution of the structure–agency question was "in the air" at that time, I think it fair to say that the field has now moved on. There were always some difficulties with it. It was extremely abstract with respect to the specific social relations entailed by structure, it emphasized the social to the neglect of the socio-ecological, and it is not clear that the separation of agency and structure was ever completely overcome. Later, postmodernism and poststructuralism would have seemingly nothing to offer. Their antihumanism meant that agency was no longer on the agenda and structure had undesirable teleological associations. And yet, interestingly, there has been a bridge to the future through the closely related actor–network theory, or ANT, in which the structure–agency question is, in effect, resolved by attributing agency to the actor–network. Through the idea of emergence it has also resonated with other ideas circulating through human geography, particularly that of complexity.

However, this has come at a cost. ANT assumes essentially a flat ontology. There is no differentiation between human and nonhuman objects, and given the utter irreducibility of every actor–network, there can be no general social processes. The confrontation with more materialist understandings of the world of the Marxist sort refuses to go away, therefore.

THE STRUCTURE–AGENCY DEBATE

There is a history of determinisms in human geography. To a large degree, their succession mapped out the course of changes in geographic thought in the 20th century. They started out as environmental, proceeded to a spatial determinism, and ended up with historical geographical materialism, which, in the eyes of some at least, was a form of social determinism. Each instance can be regarded as a set of limits or a structure to which individuals must adapt. Each can be regarded, rightly or wrongly in the case of historical geographical materialism, as marginalizing the role of human agency. Each in turn elicited a critical response attempting to achieve a greater degree of "balance" in explanatory strategies or asserting the dominance of agency over structure. It was this set of arguments that Derek Gregory (1981) termed "the structure–agency" debate.

The most well known of the determinisms is, of course, environmental, which dominated academic geography in the early part of the 20th century, many of whose claims now strike us as absurd, even ludicrous. More recently, the work carried out under the spatial–quantitative revolution was depicted by some as yet another, this time "spatial," determinism: Instead of the natural environment determining human activity, it was spatial arrangement. Subsequently, the interest in Marxism and the development of a distinctive Marxist geography led to the same sort of dismissal.[1]

Each of these found a response. The broad response to environmental determinism was possibilism, based on the idea that natural environments represented a field of possibilities from which humans could and would choose: Limits were present, since the number of possibilities was not infinite, but the possibility of choice made environments seem less limiting than they might otherwise have been. In this instance, the emphasis seems to have been on achieving a "balanced" understanding: one that acknowledged the role of both environment and human agency. Alternatively, the center of attention shifted to a culture group, which would leave its distinctive mark in the form of a landscape, the implication being that a different culture group would have left a quite different imprint. In some ways, this represented an improvement over the cruder possibilism, since at least it conveyed the idea of some sort of socially structured rather than merely physical constraint over, or facilitation of, the choices made, even if the reified concept of culture drawn on tended to vitiate this possibility. Work coming later in the 1950s that emphasized the natural environment as a cultural and technological appraisal built on this in what were intellectually more satisfying ways.

The critical response to the revival of the spatial tradition under the aegis of the spatial–quantitative revolution was to be humanistic geography. This took off in the mid-1970s. The critical emphasis here was, however, less on the spatial and more on the appropriation from mainstream economics of notions of people and society, that is, on the way in which people had been reduced through notions of rational man—not just *Homo economicus* but also, in effect, *Homo geometricus*—to something that did scant justice to them as meaning-imposing, intentional agents who made choices based on those intentions and meanings.[2] Humanistic geography introduced a very strong, almost overwhelming, sense of agency into explanation in human geography, so much so that the sort of balance sought in early rebuttals of environmental determinism was certainly in danger of being lost, if, as we will see, it was ever worth retaining anyway. In addition, the critique of the humanistic geographers found a further anathema in Marxist geography, which was castigated for its supposed economic determinism. There were retorts, particularly from the Marxist geographers, but what really stilled the flames was what seemed to be a resolution of what by then had come to be known as the individual–society debate. This came from outside human geography and from several people almost at the same time, including sociologist Anthony Giddens (1981) and philosopher Roy Bhaskar (1979, pp. 39–47). Derek Gregory (1981) and Nigel Thrift (1983) provided reviews.

Giddens's contribution was what he called "structuration theory." Bhaskar proposed a "transformational model of social action," (TMSA) but the similarities between the two versions were very strong. The fundamental idea in both was, in effect, a deconstruction of the individual-society dualism: the discovery that the two poles are, in fact, mutually conditioning. In other words, agents were always socially structured agents, while social relations could not exist unless agents reproduced and transformed them. Social relations were a necessary condition for human action. In short, the relation between agent and structure should be conceived in relational terms.

People acted the way they did in virtue of socially defined interests: interests as wage workers, wives, or wheat farmers. In so acting, they drew on still other social conditions: institutional forms, beliefs and the discourses that communicated them, various power relations, and spatial arrangement. This did not mean just limits on human action. Social conditions also enabled. There were many ways of being a wage worker or a wife: Wives in more affluent circumstances could make choices unavailable to those in poorer ones.

In drawing on those conditions, they would reproduce them. In entering an apprenticeship to become a welder, the taken-for-grantedness of that mode of recruitment was reproduced: This was the way one became a welder. Likewise, the practice of buying a house through resort to a realtor reproduced that particular mediation between buyers and sellers of housing since it meant that realtors stayed in business! On the other hand, this conception did not preclude change. Novelty could arise from the way in which, in pursuit of interests, albeit always socially defined interests, new social institutions and, therefore, practices emerged: smaller, more child-centered families, or multilocational firms, for example.[3]

Bhaskar had nothing to say about geography. Time entered in but only in terms of the reproduction and transformation of social relations as necessarily time-dependent processes. Giddens's theorization was different in that regard, and for that reason his structuration theory attracted the attention of geographers much more than was the case with Bhaskar's TMSA. Giddens emphasized the way in which the social relations, which were the reproduced and transformed conditions of action, were inevitably both geographical and temporal in character.[4] Social systems had different capacities to span space and time. For tribal societies, interaction was necessarily local, and the past that was drawn on was limited to what was communicated orally by one generation to the next. Modern societies were very different. Revolutions in transportation and communication allowed relations to be forged over very long distances: Interaction could be conditioned, therefore, by a very different geography of possibilities than was the case in tribal society. Business connections could be secured, technologies imported, and ideas transferred around the world. The same sort of radical transformation also affected the ability to access the past. Writing and then printing transformed the keeping of records and the development of that historical memory, which could form the basis, directly or indirectly through interpretation of a seemingly disinterested sort, of social practice. At least one had a better sense of what had been tried in the past, what had worked, and what had not.

One of the implications of this was a reevaluation of the nature of causality. Historically, this had been conceived in aspatial terms. Causality was understood as a purely temporal matter. In the positivist understanding of causality with its emphasis on constant conjunctures of events, X always followed by Y, and that had prevailed with the spatial–quantitative revolution, this was very clear. The prioritization of the time dimension, however, continued. Agency meant people mobilizing particular conditions but the effect simply followed that mobilization. What Giddens's work implied was that causality was, rather, a matter of time–space. This was grasped more explicitly by Massey (1992). Any process, anything linking causes and effects, like the labor process, depends on bringing things together in both space and time. It is, in short, a matter of time–space distanciation.

Not surprisingly, Giddens, particularly his *Contemporary Critique of Historical Materialism*, in which he brought many of his arguments together, attracted a good deal of attention from geographers, as I discussed in Chapter 3. But exactly how to make use of his arguments in empirical research proved to be elusive.[5] The major problem was the very high level of abstraction at which his claims were made. Even so, and whether or not Giddens's work was implicated in this—even if one had not read his work directly it was hard not to be aware of it as a result of the contributions of Gregory and Thrift—I think it is fair to say that geographic research did come to reflect an awareness of the issues of structure and agency, how agency was conditioned by structure, and how it in turn could transform those conditions. There was an increased sensitivity to the way in which social conditions could be the object of conflict and get changed as a result of that conflict. At the same time, there might be unintended consequences of action that could affect action in the future. In any case, all these arguments about how we might better understand human action in space-time found their way into the geographic literature, though just how conscious this translation was remains in doubt. Some examples of actual research projects, written up and published, will help to substantiate these claims.

Katharyne Mitchell and the "Monster" Houses of Vancouver

Mitchell (1993) focuses on the attempts of a growth coalition in Vancouver to boost growth by encouraging the inmigration of wealthy Chinese businesspeople from Hong Kong, who were then expected to invest in businesses in Vancouver. Part of the background to this were changes in Canadian immigration law, which were introduced in order to take advantage of the anxieties of Hong Kong businesspeople immediately prior to reincorporation into the People's Republic of China. The city of

Vancouver, with the support of a coalition of businesses anxious to secure the further growth of the area, was particularly aggressive in marketing the area to these would-be immigrants.

The arrival of the Chinese, though, generated some resistance of a residential sort. This was on two different sorts of ground. In the first place, there were issues of identity. The long-term Anglo-Saxon elite resented the challenge to their status coming from the Chinese presence. Not only did the Chinese use their money to move into "their" neighborhoods, so threatening their image as Anglo-Saxon, which depended on a racial exclusivity; they also altered a landscape of, for example, mock Tudor homes that had been associated with elite status in Vancouver and replaced them with so-called "monster homes." Existing houses were cleared away and replaced with structures that consumed most of their lots and looked out of place, destroying a neighborhood aesthetic. The second ground for opposition was that of housing prices. Housing prices had increased in the Vancouver area, and this was attributed to the arrival of the immigrant Chinese. This was a problem for residents, since it made it harder for their children to buy houses.

The growth coalition, anxious about maintaining the stream of Chinese immigrants with their hoped-for implications for local growth, then took up the challenge. On the one hand, it tried to disarm the opposition morally by defining it as racist. In so doing, it called on the doctrine of multiculturalism that had recently been proclaimed as policy by the Canadian government. This was the idea of equal rights under the law while respecting fundamental differences between individuals stemming from diverse cultural and "racial" backgrounds. The original target of this had been domestic issues of integrating Canada's First Nations into the body politic and also securing that of the Quebeçois. The idea that it might be used to make immigrants feel more comfortable had not been the reason for its promotion.

As for the issue of housing prices, the services of the Laurier Institute in Vancouver, a private research agency enjoying close relations with Hong Kong investors and the local growth coalition, were procured. Their conclusion was that the recent escalation had nothing to do with the arrival of the Chinese. Rather, it had to do, in the first place, with Canada's baby boom and, in the second, with local government policy in the Vancouver area, which, as a result of the imposition of impact fees, was limiting the supply of new housing.

Rachel Silvey, Women, and Spaces of Protest in Indonesia

One of the images of the creation of new international divisions of labor in the world has been that of the relocation of low-skill, low-wage employment to developing countries and a subsequent feminization of employment as young, often single women are favored targets of labor recruitment policies. They are favored in part because of the belief that they are likely to be subject to gender norms regarding appropriate behavior and, therefore, more easily managed and less prone to any form of militant action aimed at employers: in short, more easily exploitable. One of Silvey's (2003) objectives was to show how this might be a serious overgeneralization. In addition, she wanted to demonstrate the ways in which place can matter.

Her study areas were two towns in Indonesia—Jowo and Sunda—each characterized by the employment by multinationals of large numbers of single women, many of them migrant workers who lived elsewhere. Sunda was the epitome of the image conveyed by the literature: obedience to gender norms regarding how, for example, the young women's families should come first and how they should adhere to feminine norms of "dignity and refinement." Unsurprisingly, labor disputes were infrequent. Jowo, on the other hand, was a much more militant place, and the young women were an important part of that. They understood the dominant gender norms but interpreted them differently. Rather, protest on behalf of higher wages was seen as laying a more sound future for their lives as wives and mothers. What they were thinking of in particular was the welfare of the children that they eventually planned to have. In short, a particular generalization coming out of the globalization literature stood challenged. But the remaining question was: Why were there such stark contrasts between the two places?

A crucial difference turned out to be the degree to which the young women were subject to family surveillance. In Sunda fewer of the women were migratory workers. They lived with their families or, if they were migrants, with members of extended families. In Jowo it was very different. The young women came over considerable distances to work there. They lived together. They saw each other as their "family." Most significantly, they felt emancipated from the family pressures through which the norms of appropriate female behavior were communicated. This did not mean that they were not subject to gender ideology. Rather, they interpreted it, perhaps conveniently, in a different way.

Thomas Bassett and Peasant–Herder Conflicts in the Northern Ivory Coast

The focus of this study (Bassett, 1988) was on quite severe conflicts between peasants in the northern parts of the Ivory Coast—belonging to a group known as the Senufo—and Fulani herders, who have moved south into the area from the drought-prone Sahel region to the north. They have been encouraged to stay by the Ivoirean government for their own reasons. Cattle, however, are prone to trample crops. One of the responses of the Senufo had been violence designed to drive the Fulani herders away.

The government was central to the unfolding conflict. For it, the cattle of the Fulani herders were, and continue to be, seen as a means of reducing meat imports into the country. In this way, it was hoped that a balance of payments problem that had led to International Monetary Fund (IMF) loans and the austerity programs that typically go along with them could be rectified. The goal has been to sedentarize the Fulani. Given existing land tenure relations in the Ivory Coast, the government has had the power to authorize grazing rights. However, government policies have also been in part responsible for the intensity of the peasants' reaction to the presence of the Fulani. Through them, the peasants have been in the process of integration into the cash economy. As part of the government drive to industrialize, they have been encouraged to devote more land to cash crops, particularly cotton. But these then have to be sold to government marketing boards, and the cotton is bought below the world price. This leaves the immediate producers considerably poorer than they might otherwise be. The surplus that is in effect extracted from the peasantry through the pricing mechanism is then what the state draws on for its investment in processing plants (sugar, cotton, palm oil).

All might still have been well and proceeded without conflict if it had not been for ecological considerations, which throw the Fulani and the Senufo together. On the one hand, the Fulani are attracted to areas where cultivation is occurring and thus where the tsetse fly, a major scourge of cattle, is much less of a threat: In the more densely settled areas, the wild animals that are the carriers of the disease have been reduced in numbers through hunting. However, this increases the likelihood of trampling of Senufo crops, something the peasants, whose material circumstances are already reduced as a result of marketing board policies, are unlikely to take to kindly. On the other hand, Fulani herder operations are not without their attractions for the Senufo peasants, so the situation is a highly contradictory one. Wherever there are cattle there is manure, so the land grazed by the cattle, particularly the manure-rich lands close to their corrals, is highly desirable for the peasants. This, however, increases tendencies to trampling of crops and increases motivation on the part of the Senufo to chase the Fulani away.

Land rights have also been an issue. As part of the sedentarization policy, the government has invested in dipping tanks, improved pastures, and corrals. These are viewed with suspicion by the locals since under traditional tenure, while outsiders can be given temporary usufruct rights, land improvements are forbidden since it makes it harder for locals to reclaim the land at a later date. But again, it is unlikely that these concerns would have been voiced if the Fulani and the Senufo were not drawn together into an odd, if hostile, embrace.

*   *   *

By the mid-1980s, the vexed problem of structure and agency, that is, of causality in human geography, appeared to have been resolved. To what degree this resolution at the theoretical level then conditioned the way human geographers structured their explanatory accounts, or the degree to which it merely rationalized something already underway, is uncertain. The ground for the welcome given to the resolution celebrated in the Gregory (1981) and Thrift (1983) articles might be regarded as something that had been well prepared. For example, David Ley was not the only person who had sensitized geographers to agency. There was also the rediscovery of conflict by political geographers, subsequent to Julian Wolpert's initiation of studies of what would come to be known as locational conflict, and the increasing awareness of the role of social struggle entailed by the interest in Marxism. Likewise, the idea that action had its conditions, some of which might be the unintended consequences of yet other actions, was not a new one. Marxist geographers were well acquainted with Marx's famous statement about men making history.[6] But you did not have to be a Marxist geographer to understand that. Locational conflict obviously did not just happen: It had its conditions, including urban renewal and highway construction policy and the increasing self-confidence of historically oppressed peoples. Accordingly, whether or not Katharyne Mitchell, Rachel Silvey, and Tom Bassett had read Giddens or Bhaskar or Giddens's interlocuteurs in geography, Gregory and Thrift, might be beside the point. Agency and the way it was conditioned, and how those conditions were transformed through agency, were already in the air and becoming part of standard theoretical practice in research in human geography.

This standard practice is clear in each of the articles briefly discussed here. In all of the stories told people are clearly acting. They may be coming together to attract immigrants from Hong Kong and to oppose the monster houses, multinational employers, or the Fulani herders, but they are acting. Furthermore, they are *social* actors: They act as genderized women, as Canadians of an elite Anglo-Saxon heritage, as businessmen whose future growth depends on the expansion of the Vancouver economy, or as people practicing a particular way of life as herders and, increasingly, as fulfilling roles in a wider sociospatial division of labor. In other words, they cannot help but act in the ways that they do if they are to maintain a particular identity or realize interests that, again, are socially defined.

Likewise, in each of the papers, the authors are alert to the role played by the unintended consequences of actions. In the Vancouver monster house case, Canadian policy looms in the background. Without it the local growth coalition could not resort to the idea of multiculturalism as they struggled to resist the demands of the existing residents and invoked the threat that Vancouver would no longer appear a friendly destination to the immigrant Chinese, even though appearing friendly was far from the intent of the Canadian government. On the other hand, if it had not been for the imminent return of Hong Kong to the People's Republic, it is unlikely that Canadian immigration policy would have had its greedy eyes cast on wealthy Hong Kong immigrants.

In the Ivory Coast, government policies of industrialization around agro-industries and the interest in developing a domestic meat industry were important considerations. But it had not come about with a view to sedentarizing the Fulani. Only later as they fled the more risky climatic conditions of the Sahel would their value to implementing government policy become apparent.

Likewise in Indonesia, the young women's escaping the watchful eye of relatives and the enforcement of a particular interpretation of gender ideology, along with the fact of living together with other single young women, opened the door to resistance in the workplace.

STRUCTURE–AGENCY: SOME CRITICAL NOTES

> Human geography then was never quite the same again. A radical claim perhaps, but one that possibly does need to be made since it is more commonplace to attribute this sea-change to the exclusive effects of post-modernism and/or post-structuralism.
>
> A moment though it was. Lost. Forgotten. Buried. Instead, the 1990s were characterized by at times acrimonious exchanges as geography, along with the humanities and social sciences, both resisted and embraced a plethora of "new" thinkers. . . .
>
> —GREGSON (2005, p. 25)

All of what I have related so far about the structure–agency question seems convincing enough, and certainly an advance on what had gone before, at least in articulated forms, in human geography; and by no means is Gregson the only one to believe that or to lament the marginalization of what had been achieved. As evident by her quote, it was soon to be overtaken by the antihumanist claims of poststructuralism for which agency had no leverage at all. The position of the Marxists was different. They were onlookers to the debate but apart from a critical comment from Harvey[7] (1987), they did not react. In retrospect, they should have. In this section I review the standard resolution of the structure–agency debate as it stood at the opening of the 1980s first from the standpoint of historical materialist geography and then from those influenced more by "post" readings of the subject matter.

To start, a complaint at the time, as I remarked earlier, was that useful as structuration theory might be as a heuristic device, its help in conducting empirical research was actually very limited (Gregson, 1987). One could make similar comments about Bhaskar's transformational model. In retrospect, one of the reasons for this was the high level of abstraction at which social relations were defined. Exactly what sort of society was being taken into account was not part of the argument. Accordingly, particular dynamics, particular imbrications of the different moments of the social process, eluded consideration.

A second silence, of particular interest to human geographers, and not just those of a Marxist bent, concerns the incorporation of nature into the understanding. It isn't there. The material basis of social interaction goes unremarked. Geography's ecological moment in its pre-spatial–quantitative revolution days may have left much to be desired, but the relations it focused on, whether with the "natural environment" or between people as fundamentally biological, as in the concern with population issues, were, while extremely one sided, very apropos. Social relations are always ecological relations dealing with the nature of the world, including human nature as a substrate that makes society possible and that at the same time, like the natural environment, has been an object of social intervention and manipulation.

Finally, one has to register a concern that perhaps the various attempts to close the gap between agency and structure fall short of an adequate resolution. Recall that Giddens and Bhaskar in their very similar ways tackle the problem of separation of structure from agency by demonstrating their interdependence: that social structures are a necessary condition for people acting, just as the reproduction and further transformation of those structures depends on that same action. What is troubling here is the sense of the social world as a smorgasbord, albeit a necessary smorgasbord. One has to choose. Individual goals cannot be achieved without drawing on social relations that somehow enable. Likewise, there are degrees of freedom within which one has to work. Giddens is insistent that the reflexivity of individuals always makes a difference. McLennan's critical conclusion is useful:

> The main impression is rather that in the face of a barrage of literature aimed to "decenter" the human subject, Giddens (rightly or wrongly) holds on to a fairly orthodox humanist notion of the conscious, rational self. (1989, p. 210)

In this regard, there is some virtue in Althusser's view of people as, as he put it, bearers of social relations. Just by living in a capitalist world, one is enjoined to act in a manner appropriate to it. Individual interests are already social interests. The notion of freedom at the center of the concept of agency is, as Marx averred, an ideological category: not a universal in the way we experience it under capital but something given by our social relations.[8]

Postmodernism and poststructuralism might not seem to offer anything useful in this debate. In fact, they might seem entirely irrelevant. The very idea of structure suggests a teleology that is alien while an antihumanism—sometimes latent, sometimes explicit—seems to ban the idea of agency altogether. I suggest, though, that ANT, something with strong poststructural affinities, might be helpful in thinking through the problem of "how and why things happen." There are three major claims to attend to here. The first is the socially constitutive role of nonhuman objects: other organisms, natural forces, the humanly transformed bits of nonhuman matter, organic or otherwise, with which we surround ourselves or are surrounded. Second is how the coming together of objects, human and nonhuman, in networks of relations creates the effect of agency. It is the actor–network that brings about change in the world and not the individual subject or object acting, somehow, alone. Finally, every actor–network is particular. General explanations are impossible. Every actor–network is driven by processes, associations of objects particular to it. There can be no general processes to resort to as in conventional social science. Depth ontology is out.

A vivid example of how this can work is provided in Chapter 1 of Timothy Mitchell's (2002) *The Rule of Experts*. Mitchell is a political scientist and not a geographer, but a chapter in his book has clear relevance and his book overall certainly attracted the interest of geographers. In this chapter, entitled "Can the Mosquito Speak?," Mitchell's objective is to show how a severe outbreak of malaria in Egypt in 1942–1944 resists explanation in the terms of social science as we have typically understood it, mainstream or critical, and that it needs to be understood in a different light: that of the relations between a network of different objects, events, forces, human and nonhuman. Several things came together. The construction of the Aswan Dam was to have unintended consequences. Not least, the perennial irrigation that it facilitated created breeding grounds for the mosquito. It also meant that the fertility of the soil downstream was no longer replenished by the silt deposited by the annual Nile floods. Artificial fertilizers were required. Prior to World War II, this had not been a problem, but the nitrates from the Atacama Desert in Peru and Chile were by then being diverted for the production of explosives for the American war effort. All this meant an underfed Egyptian peasantry vulnerable to the ravages of malaria. To make matters worse, the Japanese invasion of the Dutch East Indies meant there was no longer any cinchona out of which to manufacture quinine, which was the major treatment for malaria at that time. In this understanding, there is no room for human agency or for social structure. The actor–network is constituted by forces that are both social, such as the war, and nonsocial, as in the palliative properties of quinine or the effects of the Aswan Dam. It is these different objects affecting one another in various ways and then acting together that makes the difference in the world.

There are some problems with this understanding. Its rejection of general social processes conditioning the concrete, the rejection of difference between objects, how they are essentially all treated as the same, human and nonhuman, suggests a radically flattened view of the world, which testifies again to its "post" affinities. In this regard, it fits comfortably within broader tendencies in contemporary human geography, including a skepticism toward the verticality of relations suggested by concepts of scale (Marston et al., 2005) and territory. This has made some others uncomfortable. In a review of the structure–agency question, Gregson (2005) has tried to assimilate some of the claims of ANT to what are now the more conventional understandings of people like Giddens. While recognizing the force of the idea of actor–networks and of the significance of the nonhuman in causal understanding, she is reluctant to let go of notions of human intentionality and how networks get reproduced through more conventional notions of human agency. So, in a discussion of the outbreak of foot-and-mouth disease in Great Britain in 2001–2002, she recognizes the crucial effect of the virus (who could not?) and the role of animals as part of a network but she then adds:

> We do not live in a flattened world of cohabitation, but one where the biographies of certain sorts of animals at least—many of them who just happened to be in the wrong contiguous locations at the wrong time—are terminated by human agency alone, and by human agents whose actions are themselves shaped by the conditions of the farming industry and by meat export markets in particular. (2005, p. 34)

Equally significant about the contribution of ANT, though, and one that has attracted less attention, is the way it emphasizes the role of emergence in how and why things happen. Timothy Mitchell refers to the way in which various objects, events, and forces "emerge together in a variety of combinations." Different things and forces interact with one another in a network of relations, and as they are changed in the process so the capacity of the network to make a difference gets changed as well. Again, this connects with recent tendencies in human geography, and it is to these that, by way of a conclusion, I wish to turn to next.

BACK TO HOW AND WHY THINGS HAPPEN
IN HUMAN GEOGRAPHY

I want to consider first the different ways in which human geographers have recently shown an interest in the phenomenon of emergence: how objects come together and in their interaction acquire new causal powers. The first example is Doreen Massey's definition of geography—not just human geography—as what she has called a "complex, historical science." By "complex and historical," she means the fact that geography's objects of interest occur in open systems where change in the entities is constant. Geography does not study systems subject to equilibrium, like those studied, erroneously or not, in economics and indeed in its offspring location theory. Geographical systems are not characterized by self-correcting negative feedbacks, though over the very short term they might be apparent, but by positive feedback mechanisms as interactions result in the acquisition of new powers. Relations in geographic systems are nonlinear. Thresholds are reached, and qualitative as well as quantitative change occurs.[9] Equally if not more significant is the way in which Massey parlays the contingent or what she prefers to call the happenstance into an understanding of emergence. The spatiotemporal is not just a contingent condition; rather, it is the source of new sorts of causal power (or liability) for the agents affected. New relations can be forged on the basis of a coming together of events and conditions.

Aside from Massey, others have also theorized along these lines. One thinks in particular of Philo's geographical history (1994): how past geographies are part of the explanation for such crucial events as the rise of capitalism, the formation of nations and states, and the production of class and race-based conflicts. There is also the interest that geographers have shown in path dependence, though this now seems to be morphing, albeit in a promising direction, into an interest in complexity theory (Martin and Sunley, 2007).

A final and earlier example, but one that is also significantly different in how it is situated, comes from Harvey. This is the idea of what he has called "a structured coherence" or a "territorial structured coherence." This emerges in the context of the attempt to secure a class compromise between capitalists and workers joined together in a geographically defined labor market:

> The class relation between capital and labor tends, under the conditions described, to produce "a structured coherence" of the economy of an urban region. At the heart of that coherence lies a particular technological mix—understood not simply as hardware but also as organizational forms—and a dominant set of social relations. Together these define models of consumption as well as of the labor process. The coherence embraces the standard

of living, the qualities and style of life, work satisfaction (or lack thereof), social hierarchies (authority structures in the workplace, status systems of consumption), and a whole set of sociological and psychological attitudes toward working, living, enjoying, entertaining, and the like. (1985b, pp. 139–140)

This is an extraordinarily interesting claim. It is clearly redolent of ideas already broached previously here. There is the idea of emergence as different relations get changed in their relation to one another. There is the idea of the uniqueness emphasized in ANT. There is also the idea of empowerment: how capitals are empowered as a class as a result of the creation of such a structure. And this, of course, is where Harvey departs from the flat ontologies that have become so popular of late in human geography.

This is not to say that there are not some affinities between the vision of historical geographical materialism and the sorts of ideas that Massey sets out in her article on geography as a "complex, historical science." In the particular position that she has adopted, there are, furthermore, some overlaps, even complementarities, with historical geographical materialism. While capital does indeed have its laws of motion, it also has a concrete trajectory of changing technologies, forms of organization, products, and geographies that recall Massey's emphasis on openness and how that openness is to be conceived as a result of the chance juxtaposition of flows and relations. In understanding the emergence of capital, one can even grant some credence to a geographical history that, while formulated by Philo, is entirely congruent with Massey's vision.[10] Likewise, the sort of spiral causality emphasized by Massey is fundamental to the accumulation process at the heart of historical materialist understandings of capital. It is also an idea drawn on in understanding more concrete expressions of the geography of accumulation. The thesis of leapfrogging developed by Storper and Walker in their (1989) work on geographically uneven development is a case in point, combining both the sort of chance origins argued by Massey with processes of a path-dependent development that is also place dependent.

There are, however, some serious limits to these overlaps and affinities, and they have to do with the way in which the accumulation process structures networks, creates new structures in order to confront its contradictions, and selects in certain aspects of the happenstance and rejects others, all the while generating something that is more than the sum of its parts. Uncertainty is not something capital is passive about. Rather, it tries to anticipate events and to control them so as to limit surprises. Much of what comes together in a place is intentionally *brought* together, assembled. Capital intervenes and pulls in the distant, whether it is ideas about organizing production from Japan or folk remedies from around the world for the pharmaceutical industry or labor from Mexico.

Most problematic of all with flat ontologies is the radical openness to the future that they assume. Here, of course, we return to the problem of essences. There is an element of chance in determining where capitalism first emerged and when, but I am going to conclude by arguing that these events have to be seen against a background of enabling conditions that owe nothing to chance. There are enduring tendencies in human history, despite what the postmodernists say. I would argue that the following are uncontroversial:

1. That people's ability to produce has increased over time, almost monotonically; the increase has been very rapid since the advent of capitalism but feudal England was light-years ahead of the Stone Age and even the Neolithic.
2. That this has been due to an increasing socialization of production, notably through the development of the division of labor.
3. The development of the division of labor has, in turn, entailed the development of exchange and the means of exchange, in which money has been preeminent.
4. The increase in the ability of people to produce has opened up the prospect of producing a surplus and therefore of a class society in which some produce the surplus and others, an exploiting class, take possession of it for their own purposes.

In other words, by the late Middle Ages, a number of the conditions for the transition to capitalism are already in place, including markets, if not yet in labor and land, and a marked degree of social differentiation, including access to the means of physical force, as a result of the class relation between lord and serf, and money. More briefly still, if capitalism was an accident, it was an accident waiting to happen.

This is to concentrate on the emergence of capitalism. Once capitalism is up and running, then chance continues to play a role. Just how this occurs bears careful examination. Capitalism is an expansionary system of production. Firms have to grow. The intensity of competition provides them with no alternative if they are to avoid going out of business. Capitalist development has been, from one standpoint, a succession of innovations that bring together all moments of production: not just technical but also institutional, geographic, the employer–employee relation, the discursive, and so forth. There is something unpredictable about these. They can be explained after the fact and Massey can be regarded as pointing suggestively in that direction: the role of random juxtapositions of institutional forms, technical innovation, and ideas. They assume forms of lock-in of the sort associated with path-dependent development. And they can revolutionize the landscape, as in the low-density suburb and the new industrial districts associated with new, emergent sectors, or at more microlevels: for example, the pattern of gentrification in a particular city. It is in these terms that we can imagine the idea of possible geographies. Yet they remain *capitalist* geographies, since without the conditions, the incentive framework laid down by capital, they could not possibly exist.

NOTES

1. A notable, perhaps notorious, contribution to this argument was that of Duncan and Ley (1982).
2. It is not true to say that spatial–quantitative geographers were always unaware of this issue. One solution proffered by, among others, Curry and Hägerstrand was that of probabilistic modeling. See Chapter 2 of the current volume.
3. As Bhaskar emphasized, this resolution of the structure-agency problem needs to be sharply differentiated from one offered earlier by Berger and Luckmann (1967). This is the claim that people construct society and only then does it condition human action. Bhaskar's point is that people never act except through social relations which pre-existed them.
4. This formulation was developed further by Soja (1989, p. 129) in the form of what he called the sociospatial dialectic.
5. For an assessment of the possibilities, see Gregson (1987).
6. "Men make history. But not of their own free will; not under circumstances they have chosen but under the given and inherited circumstances with which they are directly confronted." (Marx, 1869, p. 146).
7. "I see absolutely no point in discarding the fundamental Marxist conception, that people—both individually and collectively—are perpetually struggling so as to make their own history, but that they do so under material conditions that are given not made by them, in favor of some looser and vaguer language of, say, agency and structure, however safer because politically innocuous the latter might be" (1987, p. 368).
8. "The sphere of circulation or commodity exchange, within whose boundaries the sale and purchase of labor-power goes on, is in fact, a very Eden of the innate rights of man. It is the exclusive realm of Freedom, Equality, Property and Bentham. Freedom, because both buyer and seller of a commodity, let us say of labor-power, are determined only by their own free will. They contract as free persons, who are equal before the law. Their contract is the final result in which their joint will finds a common legal expression. . . . When we leave this sphere of simple circulation or the exchange of commodities, which provides the 'free-trader vulgaris' with his views, his concepts and the standard by which he judges the society of capital and wage-labor, a certain change takes place, or so it appears in the physiognomy of our dramatis personae. He who was previously the money-owner now strides out in front as a capitalist; the possessor of labor-power follows as his worker. The one

smirks self-importantly and is intent on business; the other is timid and holds back, like someone who has brought his own hide to market and now has nothing else to expect but—a tanning" (*Capital* Volume 1, p. 280).

9. An example is provided by Hägerstrand's model of migration. The initial nucleus of migrants reaches a certain point, and beyond that point a "home away from home" (travel agents, newspapers, mosques, social organizations, food availability) is created to give further migration a much stronger impetus.

10. Thus, one might regard capitalism as an accident rooted in various events that came together in the Tudor period: events that hastened the creation of markets in labor and in land. I am thinking of events like the dissolution of the monasteries and the sale of the land to court favorites while at the same time suddenly leaving large numbers of monks and nuns without access to means of subsistence. Another would be the rise of the wool trade. "The sheep doth eat men," complained Thomas More in *Utopia*; a tract devoted to understanding why, during the Tudor period, those who thought they had hereditary rights to the land—rights that they could not sell—now found that through main force they didn't and that they were being replaced by the sheep of engrossing feudal lords.

CHAPTER 8

Making Space for Human Geography in the Social Sciences

Geography's anxieties about itself are of long standing. In Peter Taylor's (1985, p. 93) words, the field's "practitioners have exhibited a perpetual lack of self-confidence." Explaining itself to the rest of the world has been a continuing problem, though this is partly because of the remarkable diversity of its subject matter. It includes both physical and human geography, and so calls on a huge diversity of processes and specialized knowledges. One consequence is that it lacks the legibility of, for example, economics or political science or even sociology, and this is true not just for lay audiences but for academic ones too. In the United States in particular it suffered for a long time from a sense that its position in the academy was a relatively precarious one. Symbolic of American precariousness is the fact that the field is represented in only one of the prestigious Ivy League schools.[1] It also has no representation in many of the other great American universities like Chicago or Stanford. In Great Britain, on the other hand, geography has a secure place at both Oxford and Cambridge. Part of the difference is the relative strength of the field at the high school level, and this is also true in Canada, Australia, and New Zealand. In that regard the United States is an outlier. On the other hand, even in these more secure sites, geography has had its detractors.[2]

One result has been a continual navel gazing trying to find some disciplinary fix. Yet the field's own diagnoses of its dilemmas have often been unhelpful. One thought in the earlier years of the last

199

decade was that its problems stemmed from not having an object of study like other fields. Economics had the economy, firms, markets; sociology had society, family, institutions; while political science had the state and everything to do with it. Geography's response was the region, which in retrospect was not a great success. In the same vein has been the thesis that human geography has been parasitic on the other human sciences[3]; that human geography, in fact geography altogether, has no theory of its own but must rely on other fields. This, it is argued, helps to explain the strength of the centrifugal forces in human geography (Johnston, 2005).

The latter claim is particularly interesting since it demeans the recentering forces in the field, what it specifically has to offer as a set of abstract concepts. In this regard, it is useful to note that what the human geographer makes available to the rest of the academy simply will not go away. The other social sciences always draw on geographic knowledge and theory in some way, often imperfectly, just as human geography draws on them. Sociologists have a long tradition of community studies in which the implicit assumption is that living in close proximity counts for something. In political science comparative politics is a major subdiscipline. To practice it one needs to know something of the specificity of countries. Whether the geographic imagination gets adequately represented in these endeavors is something else. Most telling of all, perhaps, when university geography departments are abolished, they often reappear in another form: Centers for Urban Studies, Area Studies Centers, Centers for Environmental Studies or even Forestry and Environmental Studies, though the very idea of "center" is significant in itself, suggesting that geography does not benefit from the current organization of the university and thrives best when in close cooperation with other fields. As I outline in this chapter, in such a context human geography has as much to teach as it does to learn.

Second, we should note how the fact of space and space relations subverts fundamental assumptions in social theory. Microeconomic theory builds itself on the notion of perfect competition; but there can be no perfect competition when the literal spacing of firms provides some market power. Political science's master concept is the state. But any notion of sovereign power is compromised by spatial difference and the way it has necessarily to be incorporated into state structure and acknowledged in the practice of the state's central branches. Some political scientists have worried about why the world is divided into many states. The fact of globalization and the mobility of capital has suggested to them that territorial division might be a hangover from precapitalist days (Callinicos, 2004), even while these tendencies coexist, and necessarily (Harvey, 1985a) with ones of reterritorialization. In short, social relations are always and necessarily spatial ones. Social processes unfold over space and incorporate a spatial moment. Space is not just a correlate; it is actively mobilized in securing social ends, and it can limit what is possible, suggesting that there can be no defensible social theory that does not recognize the significance of geography.

Yet in their more abstract claims as opposed to their empirical work, it is true that the other social sciences have, and with some important qualifications, given human geography short shrift. As many have noted before, they have tended to emphasize time rather than space. Economists have their theories of economic growth emphasizing, *inter alia*, its preconditions, changes in the division of labor over time, and the creation of virtuous circles between investment, consumption, and more investment. Sociologists long put at the center of their grand theory the idea of modernization: the displacement of more traditional societies by modern societies and how that process proceeded—through time, of course. Similar arguments could be made about political science's metanarratives: a prioritization of time over space.

This is one problem, but not the only one. Another that gets in the way of what human geography has to offer are the universalizing ambitions of other social sciences. From that standpoint, geography is utterly destabilizing. Human geography has certainly had its own self-conscious universalizing moments, notably during the spatial–quantitative revolution, when the emphasis on a quantified relative space allowed some convergence with urban and regional economics. Regional difference, however, would not go away. The tendency was to try to hold it constant through appeal to what were often the black boxes of dummy variables or covariance analysis. In this way the effects of space could be revealed in all their orderliness. This could never work. Rather, universality has to be seen in some sort of productive relation with the particular. They cannot be separated in these ways, and to say that they are mutually dependent is inadequate. They incorporate each other; the relationality of their relationship is of the essence. We might try to define the social process in universal terms, but geography is an ineluctable part of it, imposing its own constraints as well as opening up new possibilities.

To the extent that they acknowledge geography, the human sciences have tended in two distinct directions. The first is indeed to incorporate it in their own universalizing way. Urban economics and urban sociology are classic examples. The new economic geography of people like Paul Krugman and Edward Glaeser is a continuation of this approach. This has been remarked already by geographers and is taken up in more detail later in the chapter. Less noted, less subject to critique, has been the embrace of comparative studies. Both political science and sociology have a long tradition of this, and there are a number of methodological works devoted to honing analytic skills in that direction (Ragin, 1987; Tilly, 1984). But again, we are back to "holding things constant" and sometimes, as in quantitative studies drawing on multiple countries, in a very obvious sense and uncluttered by concerns about the difference that geographic specificity might make. Geography becomes a problem that has to be neutralized rather than drawn on creatively.

Second, to the extent that other social sciences are self-conscious about human geography and what it might contribute, the tendency is often to a very outdated, pre-1960s vision of the field. As we have seen, this was one dominated by simplistic notions of people–nature relations devoid of any underpinning in terms of social relations and the particular relations with nature that they tended to entail. The economists have been to the fore here. Some of the titles of their papers give the idea: "Tropics, Germs and Crops: How Endowments Influence Economic Development"; "Natural Resource Abundance and Economic Growth"; and "Geography, Demography and Economic Growth in Africa." The tropics and their "natural" challenges have evidently been an important theme. Equally unhelpful has been the "pop" social science of the media. Ideas of "flattening the world" (Friedman, 2007) or "the revenge of geography" (Kaplan, 2012) are redolent of geographic understandings that now seem disingenuous or utterly archaic, a throwback in the latter case to James Fairgrieve's (1915) *Geography and World Power.*

There are important exceptions to these claims: writings that express a keen spatial imagination. Most of these contributors are sociologists or political scientists. Among the former, one would have to acknowledge the work of Anthony Giddens, referred to earlier in the book, Immanuel Wallerstein, Michael Mann, Michael Burawoy, and Andre Gunder Frank (1972), who wrote a wonderful critique of modernization theory and the enclosed spaces that it assumed. The political scientists of note include Timothy Mitchell, whose work was referred to in Chapter 7 and James C Scott. The interest in globalization across the social sciences also produced some notable contributions from nongeographers, including Hirst and Thompson's (1996) *Globalization in Question* and the various writings of Robert Wade. There are others that are more difficult to position with respect to discipline, like Edward Said, who through his arguments about Orientalism, has had a major impact on the field. Mike Davis also deserves mention. All of these are people with whom geographers have been able to engage constructively.

These remain, though, exceptions. Given the development of human geography since the 1960s, the insights, and the strong ideas that it has thrown out, the other social sciences should be able to do better than they have. Just how and what human geography can contribute to them is the focus of the rest of this chapter.

HUMAN GEOGRAPHY'S CONTRIBUTION

In General

Human geography's contribution ultimately goes back to its key concepts. Regarding these, there seems to be general agreement that three predominate. These reflect the history of the field, how it has developed as part of the division of labor of the social sciences, as much as any more theoretically defensible breakdown: space; place, including geographical differentiation; and the environmental. Quoting Massey (1999a), "For us, human geography's take on the world derives from its standing on the ground of the triad space–place–nature" (p. 4). Of these space, or more accurately space–time, exists at a higher level abstraction than the other two. To talk about the natural environment only makes sense in terms of the idea of "environing," just as place involves "placing."

At lower levels of abstraction again are still other concepts that are commonly drawn on in human geography. These would have to include the following, in no particular order and with no pretensions to inclusiveness:

• *Geographic scale.* This is fundamentally a way of thinking about space relations and can apply both to places (neighborhoods through regions to countries) and to natural environments. Prior to the spatial–quantitative revolution, it was common to think in terms of environmental factors at different geographic scales: how climate typically varied much more gradually over space than geology, for example, though there was also the distinction between macro- and microclimate.

• *Location.* This is commonly associated with the spatial–quantitative revolution and location theory. It also entails an emphasis on movement since there can be no location without it. And without movement there can be no urbanization or dispersion. It sits uncomfortably in an emphasis on places and the sense of stasis, though the new regional geography (Chapter 5) clearly challenges this equation.

• *Networks.* Movements make no sense if they are not channeled in some way. Again, this is a concept that attracts attention in spatial–quantitative geography largely in the context of transport, though Hägerstrand's work on communication networks showed that even then its meaning was considerably broader. The recent interest in flat ontologies (Chapter 7) has led to a revival of interest.

• *Settlement.* One no longer hears much about this. It seemed to lose traction with the spatial–quantitative revolution, perhaps because of its static, undynamic connotations. "Location" became the preferred term. It has senses that remain useful, notably that of relative immobility: a necessary counterpart to movement, therefore, and sometimes contradictory to it, as Harvey argued in his work on the geopolitics of capitalism.

• *Territory*. This is derivative of exactly this contradiction: the desire to protect something immobile against the challenges of a wider, shifting geography and the movements it engenders. Recently, perhaps in the context of globalization, it has acquired a processual sense: processes of territorialization and deterritorialization.

• *Landscape*. This enjoys its closest affinities with place and with environment, including built ones and allowing for the skepticism with which the "natural" in natural environments is now viewed. The idea of place-lessness and homogeneity in built environments has reaffirmed its close connections with human geography's master concept: space.

Specific Contributions

A crucial critical function in the relation between human geography and the other social sciences is what Harvey (2005a) has termed "geographical deconstruction." No social science research lacks at least an implicit geography, even if it is conceiving of the world as situated on the head of the proverbial pin. So, as Harvey has suggested, we need to examine that research from the standpoint of the central geographical concepts, how they are represented (or not), and how they are conceived in relation to the social or socioecological process.

The work of geographers on scalar relations is symptomatic of the possibilities. The way in which more global forces of a material or discursive form engage with difference and vice versa was clearly apparent in some of the cases reviewed in Chapter 7 and provides a strong antidote to those easy notions of homogenization that, while apparent in their most outrageous form in "pop" treatments of the question,[4] have taken a long while to expire in the more serious social science literature. In the Ivory Coast it was a question of how global discourses and IMF pressures encountered a very distinctive configuration of socioecological forces. In Indonesia, conceptions of gender relations and the politics of family varied from one workplace to another and provided very different challenges to the employment relations of multinational firms. One should also mention Mark Boyle's (2001) critique of the social science literature on diasporic nationalism. This is because it raises very similar questions of the importance of local context and, in this case, those contexts in which different diasporic streams from the same origin land up.

On the other hand, the work of Susan Smith on the Peebles case discussed in Chapter 4 or studies on the politics of scale more generally emphasize the relative nature of the local–global relation rather than the absolute form, often confused with the international, that is most often assumed in the globalization literature. As a different sort of antidote to the received wisdom on globalization, Michael Storper (1987) has drawn attention to how the global gets made from particular local developments, how diffusion of institutional forms or modes of organizing production makes the global.

Geographical deconstruction, though, is but one expression of the idea of "spatialization." Spatialization, of both theory and method, has been a dominant theme in the geographic literature since the spatial–quantitative revolution of the 1960s. It has clearly changed its form over time. It meant something different in the 1960s than after the engagement with social theory of a more critical kind. Location theory represented a spatialization of sorts and one that could be built on, as indeed it was, but its atomistic assumptions always imposed strict limits on what was possible: the inclusion of new "variables," for example. Today the dominant meaning is quite different. Abstractly, I would argue that it is to understand some object, practice, or structure of relations—the state or some ethnic minority, for example—in terms of its relations over space with other objects, structures, practices, the internal spatial relations significant to its functioning, reproduction, and transformation as well as connections between its internal parts and some exterior. These relations can be of a material sort, of an ecological or value character, for example, and also communicative, as in some discursive formation, the circulation of meanings, though these almost certainly go together: Value communicates through the famous "price signals," just as shared meanings underpin material exchanges.

What I am thinking of here bears close resemblances to ideas put forward by Harvey and Massey. Harvey (1996) has talked about what he has called "permanences": objects, concepts, that provide a sense of thing-like self-sufficiency but that are historically transitory and that depend for their reproduction on relations with other objects and ideas. Movements, material and communicative, became primary and "permanences" their secondary expression. There are strong echoes of this in Massey's claims about how juxtapositions in space–time of influences and conditions lay the basis for the construction of new ideas and new material forms; this is very clear in her *World City* (2007) book and earlier in *Rethinking the Region* (Allen et al., 1998). Again, and what ties these arguments together closely, is the significance that both assign to tensions and contradictions resulting from these relations in space–time and that have a potentially transformative effect.

One way of thinking this through is in terms of the different moments of the social process and how these have, in fact, been given some geographical form in the work of geographers. There are some obvious examples here: Massey, Scott, and Walker on the division of labor; Agnew and Allen on power; Gough, Rose and Martin, and Wills on class relations and not to mention Harvey again, of course; Derek Gregory and Don Mitchell on discourse; and Michael Watts on the relation to nature. Table 8.1 provides other examples.

**TABLE 8.1. Spatializing the Social Process:
Central Questions and Exemplars**

The social process	Some central questions	Exemplars
The division of labor	• How is the division of labor, both as a technical and social division, projected across space? • How might one account for its geographies? • How does it contribute to the constitution of geographically uneven development?	Clark (1981); Massey (1979, 1984); Scott (1982b); Walker (1985)
Power	• How is power translated over space? • How is it constituted spatially? • How should the idea of state sovereignty be conceived?	Agnew (2005); Allen (2003); Sayer (2004)
The relation to nature	• What is the geography of technology and what are the mechanisms of, and barriers to, its diffusion? • How do in infectious diseases spread? • What does it mean to talk about the social construction of nature?	Brown (1968, 1981); Castree and Braun (2001); Cliff, Haggett, and Ord (1986); Hägerstrand (1965, 1967); Gertler (1997)
Meanings	• How is geography incorporated into discourse? • What does landscape communicate? • What is the role of discourse in geopolitics? • How do meanings get circulated?	Mitchell, D. (1996, 2000); Tuathail (1996)
Class relations	• How and why does the class relation vary over space? How do space relations condition class relations? • How is place specificity formative of class relations?	Gough (1991, 2002); Martin, Sunley, and Wills (1996); Rose (1987); Thrift and Williams (1987); Walker (1985)
Institutions	• How might one critique regulation theory from the standpoint of its implicit geographic assumptions? • What is the geography of particular institutions like science parks, universities, hospitals? • How is the state organized spatially?	Massey (1995); Sayer (1989b); Christopherson (1993, 2002); Johnston (2002); Brenner (2004)

As a result of this, human geography has been the source of what Harvey has called "strong ideas." Many of these come readily to mind: Massey's (1979, 1984) spatial divisions of labor, Berry's (1967) extension of central place theory to the intraurban case, Walker's (1978a, 1981) suburban solution, Werner's (1968) delta-wye transformation in transport networks, Philo's (1994) geographical history, Hägerstrand's (1970) time geography, Agnew's (1994) territorial trap, Swyngedouw's (1997) glocalization, McDowell's (2004) microgeographies of gender, Allen's (2003) geographies of power, Storper and Walker's (1989) "leapfrogging," and Gregory's (2004) colonial present. Many of these "strong ideas" have been referred to earlier in the book: Watts's (1983b) "natural hazards"; Scott's (1985) work on transaction costs and agglomeration; Peter Taylor's (1971) ideas about shape and its effects; and Neil Smith's (1992) "jumping scales."

This history of spatialization has affected method. Methods have to acknowledge the irreducibly spatial character of social reality, though it took some while for this to be realized. This starts with the careful specification of concepts. Many of these, like segregation, localization, or scale, are traps for the unaware and can give research a false turn. In many instances, conceptual specification moved or has moved in tandem with devising quantitative measures. This was particularly the case during the spatial–quantitative revolution. Just as space disables social theory, so too can it undermine the most basic assumptions of quantitative method. In a geographical world, the independent observations demanded by inferential statistics make no sense at all (Gould, 1970). Spatial differentiation complicates the problem of interpreting coefficients of various sorts for geographically defined populations. Rather, they have to be seen as averages. In an obvious case, a weak regression coefficient can be the outcome of a very positive one across observations in one part of the study area counteracted by a very negative one in the other segment. Areally based data are typically available only for jurisdictional units, like counties or states. But these almost invariably vary in their areas, and this can lay down one more snare to catch the unaware. This is the modifiable area unit problem. Different partitions of the same area, different combinations of observation units perhaps, are going to generate very different summary statistics. Shape also plays into it, as any master of the gerrymander would recognize; change the shapes and you change the summary coefficients.

On the other hand, the spatial imagination has in its turn contributed to quantitative methods, making them more fit for the study of a world in the production of which space is a necessary moment, something reviewed in the case studies that follow. In some regards, this spatialization of method is unique to the quantitative school. In what one might call human geography's postpositivist phase, the relation between theory and method has been more organic. Theories are developed and concepts are refined and then evaluated, drawing on archival investigations or semi-structured interviews in which the spatial itself forms an obvious and unproblematic moment allowing, by way of example, the definition of causal structures that are sociospatial rather than simply spatial. During the heyday of spatial–quantitative geography, though, it was otherwise. A law-seeking focus meant an appropriation of statistical method that was initially quite unproblematic. Law seeking, moreover, was rigidly focused on the explanatory significance of space, as indicated by the magnitude of gravity model exponents or rules of spatial arrangement, and this too served to curb the spatial imagination.

What all this amounts to, and in sum, is that social theories are only as good as the degree to which and the way in which they recognize the disturbing effects of space—of geographic differentiation, scalar relations, territoriality, the stretching of ecological relations over space. But by the same token, spatialization can only be as good as the (social) theory being spatialized. Even where there *is* a spatialization, it may be in question because of the underlying theory and assumptions about the world. The attempts by economists to incorporate what they call "geography" provide an interesting illustration of this, particularly in work on geographically uneven development rather than "the new economic geography," which is discussed later in the chapter.

The work of Jeffrey Sachs (Sachs, 2003; Sachs, Mellinger, and Gallup, 2001) has rightly attracted attention. Focusing on the question of variations in national wealth, why some countries are rich and some poor, "geography" as he defines it looms large in his explanation. This takes shape in two particular forms. The first is location: The economies of coastal areas and those with access to navigable water typically do better than landlocked areas. The second is climate: Tropical areas are problematic for economic development because of lower agricultural productivity as a result of the fragility of tropical soils, and the way in which the prevalence of infectious diseases limits human productivity. There are also some more subtle long-term effects. High infant mortality provides an incentive for parents, anxious for support in their old age, to have more children, but this means that less can be spent on the education of each child, further holding back productivity. And, of course, the correlations between these geographic "variables" and measures of development give these claims some credibility.

To add to the overall effect, they have given rise to further debate, typical of the economics mainstream, as to the independent effect that "geography" in these forms has, as for example in terms of the magnitude of a regression coefficient in a multiple regression equation. One argument has been that with respect to levels of development, the effects of "geography" are indirect and that they work through institutions.

Accordingly, colonialism operated differently in tropical as opposed to temperate climates. Climate affected the numerical presence of whites, and this affected the sorts of institutions established, which in turn affected possibilities of economic development.[5] Indeed, the regression equations confirm the mediating role of institutions.[6]

Aside from the fact that the conception of what geography means in this literature seems to be a throwback to much earlier work in the field, in particular the idea of geography as the natural environment including accessibility relations, what leaps out is the atomistic conception of the social process at work here: the world as a set of things or forces that relate to one another in a pluralistic fashion and all blithely taken for granted. Theoretical development becomes a matter of cumulative generalization, and the cause of the latter is to be advanced by new empirical studies incorporating new variables and new examinations of how they relate to one another. Ideas about context and relating the points in the statistical space that is the object of correlation analysis to those contexts and their very variable constitution are nowhere to be found. Equally, the idea of space as separate from society is like some evolutionary throwback, as if navigability could be isolated from the economics of port construction and that of the container shippers. To accept "countries" as meaningful data-collecting units is also problematic: "Countries" may have access to navigable water but their more developed parts may be a long way away.[7]

CASE STUDIES

The New Economic Geography

As well as a new regional geography, there is also a new economic geography—much newer than the one whose roots go back to the critique of the spatial–quantitative work and that I reviewed in Chapter 3. This time it is something that is being defined outside the field of human geography, specifically by economists. It is significant, though, and for several reasons. In the first place, it incorporates aspects of the difference that space makes that geographers would acknowledge as necessary. It recognizes the role of space in undermining ideas of perfect competition. Agglomeration and path dependence occupy major positions in its theorizations. Chance juxtaposition, particularly in the form of what have been called "historical legacies," is also significant. In the second place, it provides a challenge to the work of geographers, particularly economic geographers. Some of the claims made about the need for economists to get involved in the study of economic geography have been patronizing and even disdainful. Economic geographers have not been slow to respond. It therefore provides an opportunity not only to illustrate the difference that space makes but also to throw further light on the difference that specifically human geographers can make in addressing the sorts of problem that economists have recently started examining.

First, we should note the distinctive point of departure for the new economic geography, or what some geographers have termed "geographical economics." This is trade theory. This is unusual and makes a definite contribution. There is a paradox. On the one hand, classical trade theory based on the theory of comparative advantage predicts that trade should occur between countries with very different factor endowments. In consequence, one can imagine countries that are characterized by skilled labor and are capital rich engaging in trade with countries of largely unskilled labor, which are capital poor: capital- and skills-intensive products for labor-intensive, low-skill products or, for example, computers for garments. However, what we find is that countries with very similar factor endowments dominate world trade. Furthermore, a lot of that trade is not computers for garments, which would be interindustry trade; rather, it is *intra*industry, as in the trade of Scottish whiskey for French wines or German Audis for French Peugeots.

Second, the new economic geography forges a connection between geographic scales. Trade is registered as *inter*national and certainly contributes to state revenues and a country's balance of payments. But it is also about regional development. Exports are produced in particular areas so trade theory has to attend to that. Given the paradox noted previously, what will distinguish competing regions is not their costs of production; but the distinctive nature of their products. The new economic geography, therefore, focuses on product competition and not on cost competition. Trade is a way of expanding markets and allowing firms to take advantage of the increasing returns that come from operating at larger scales. "Scale" here refers both to the economies of scale internal to particular firms and to the economies that derive from their connections with other firms as a result of agglomeration: notably shared labor markets allowing a concentration of labor reserves and enhanced specialization among suppliers.

Just what accounts for the specialization of firms in particular areas is, according to the new economic geography, a historical accident. There might be competitors elsewhere, but small differences can result in the advantage being given to one rather than the other, after which economies of scale, both internal and external, take over, creating strong tendencies to path dependence. As Martin and Sunley observe:

> An economy's form is determined by contingency, path dependence, and the initial conditions set by history and accident. Forward and backward linkages mean that once an initial regional advantage is established it may become cumulative. There is therefore no automatic tendency toward an

optimal solution, as apparently "irrational" economic distributions may be "locked in" through increasing returns. (1996, p. 264)

With the lowering of costs of transportation allowing access to more distant markets, economies of scale can be exploited still further, driving competitors elsewhere out of business and producing marked core–periphery effects in the subsequent economic geography.

What exactly should geographers make of all this? There have been serious criticisms.[8] For a start, the significance of economies of agglomeration is nothing new. This is ground that has been well worked over by geographers. The work of Scott and Storper provides examples of this. Their understanding of them is more expansive than that of the new economic geographers, who tend to adhere to those that can be given some quantitative expression. Storper's (1997a) work on untraded interdependencies and what he calls the product-based technological learning characteristic of industrial districts provides important cases in point.

Second, and relating once more to the issue of what can and cannot be modeled, there is the altogether reasonable complaint that the claims set forth by Krugman and others fail to situate agglomeration tendencies and exploitation of their potential with respect to what Martin has called "the social, institutional, cultural and political *embeddedness* of local and regional economies" (1999, p. 75). Firms have to secure connections with banks that are familiar with their line of business. Connections with local universities can be significant both as vehicles for collaboration in research and development as well as in the supply of appropriate skills. Local representatives in national legislatures work to protect the economic base of their respective constituencies so that the future of a particular agglomeration can be secured. There are tacit understandings with local politicians facilitating the "right" land use decisions and interventions in local housing markets.

Third, it does not take much experience of the work of geographers in economic geography to realize that the sort of durability attributed to agglomerations in the new economic geography is grossly exaggerated.[9] There are *dis*economies of agglomeration; and while local governments may work to resolve these through highway building programs and facilitating the expansion of the local airport and housing stock, they can be very stubborn. One of the responses of firms has been through the creation of the sorts of new spatial divisions of labor envisaged by Massey. Agglomerations can hollow out, leaving behind the headquarters and research and development and perhaps the more advanced of the manufacturing processes. Apart from that, the sectoral emphasis of an economy can change. New sectors arrive on the scene in new locations. Old sectors decline along with the locations in which they were centered.

In other words, the contributions of economic geographers as opposed to the *new* economic geography coming out of economics departments are not to be disparaged. There is an accumulated knowledge there that means a reluctance to engage in overgeneralization and the sorts of simplification often necessitated in economic modeling. The requirements of mathematization put further restrictions on what can and cannot be taken into account, and the debates in human geography around the possibilities and limits of quantitative methods have induced an awareness of just how critical those omissions can be. On the other hand, and anticipating the next discussion, nor should we ignore the possibilities.

Developments in Quantitative Geography

> I explore the recent and potentially powerful movement within spatial analysis where the focus of attention is on identifying and understanding differences across space rather than similarities. The movement encompasses the dissection of global statistics into their local constituents; the concentration on local exceptions rather than the search for global regularities; and the production of local or mappable statistics rather than "whole-map" values. This trend is important not only because it brings issues of space to the fore in analytical methods but also because it refutes the rather naive criticism that quantitative geography is unduly concerned with the search for global generalities and "laws."
>
> —FOTHERINGHAM (1981b, p. 88)

In what has become the human geography mainstream, the image of spatial–quantitative analysis is indeed one of the "search for global generalities." It might seem, therefore, that notions of context and thus of particularity are inappropriate to that particular genre of research. This, however, is quite untrue. Even during the heyday of spatial–quantitative geography, Massey's (1999b) image of geography as "complex historical science" was always apparent in the work of a few, in particular Curry and Hägerstrand, as I tried to emphasize earlier. There was also work going back to the early 1970s that is in sharp refutation of the idea of insensitivity to local variation.

An early forerunner of the sort of work that Fotheringham highlights in his 1997 review was that of Casetti (1972; Casetti and Semple, 1969; Odland, Casetti, and King, 1973) on what he called "the expansion method." The fundamental idea was simple but powerful. The coefficients in a regression equation, the intercept and slope coefficients, are averages for the study area. Might it not be that they vary in some way over the study area and in a regular manner? For example, we know that the diffusion of innovations like hybrid corn or tractors is often outward from a particularly favored center; we also know that the trend of diffusion once it "takes off" in a particular place conforms to a logistic growth curve. So why not set the coefficients in the logistic growth equation as functions of distance from the growth center to show how the parameters vary over space (Casetti and Semple, 1969)?

That regression coefficients do vary over space in quite regular fashion was demonstrated by Fotheringham (1981b) in a study of distance exponents from gravity models applied to interurban air passenger flows from various points of origin in the United States. Exponents across much of the west and south were considerably higher than those for midwestern or northeastern locations (Figure 8.1). He then showed how there was some regularity in this pattern in that exponents became more negative (i.e., steeper) with decreased accessibility to the population of the United States taken *in toto*. He could, of course, have expanded a gravity model incorporating all the data that went into calculating the models for individual points of origin as a function of that measure of accessibility. But his point—that spatial structure affects long-distance travel behavior—was demonstrated regardless.

This sort of study along with the earlier ones of Casetti might seem not that far from the standard image of spatial–quantitative geography as

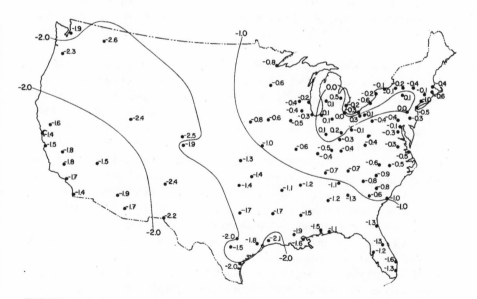

FIGURE 8.1. A map of distance exponents, air traffic flows, 1970. Each number represents the distance exponent on a gravity model for flows between respective points and all other cities. From Fotheringham (1981a), p. 426. Reprinted by permission of Taylor & Francis Ltd, http://www.tandf.co.uk/journals.

in enthusiastic pursuit of spatial regularity. It did, however, represent an evolution in a direction that would eventually be much more sensitive to spatial context. In retrospect, a crucial turning point was Rushton's 1969 article, in which he contrasted what he called "behavior in space" with "spatial behavior." Behavior in space, he argued, incorporated particular aspects of the space in which it had been observed. Journey-to-shop patterns tended to vary according to patterns of accessibility, much as Fotheringham would later discover in examining air passenger flows. For a human geography that was intent on developing general theory, this was not a good basis. Rather, he argued, attention should focus on "spatial behavior," or "the rules by which alternative locations are evaluated and choices consequently made."[10]

This is not to denigrate Rushton's project of so many years ago. Universal claims about society and space are one of the major contributions that geographers make to the human sciences. Since then, though, and as Fotheringham discusses in the survey article from which the quote at the head of this section was taken, spatial–quantitative geography has undergone some shift in its priorities. The search for spatial regularity, both that to be deduced by Rushton and that to be induced by people like Casetti and indeed Fotheringham in his work on gravity coefficients is now deemphasized, and this has allowed a clearer interest in the role of context to come to the fore. Here I would like to draw attention to three developments, all of which are picked up on in Fotheringham's survey:

1. *Spatial autocorrelation.* Tobler's (1970) law signaled not just substantive relations between places that tended to be enhanced the closer they were together geographically, but also relations between the attributes of places: the fact that among randomly chosen pairs of places, closer places tended to be more similar. For many years, this was regarded as a major technical problem, since it meant that a fundamental assumption of regression—the independence of observations one from another—was not satisfied.[11] Accordingly, the practice developed of computing measures of spatial autocorrelation for the study area as a whole in order to check to see how serious a problem it might be in a particular instance. This persists but alongside it has developed an interest in how the magnitude of spatial autocorrelation varies over a study area, so generating maps of spatial association (Anselin, 1995).

2. *Geographically weighted regression.* The same observations apply to regression. Regression coefficients have typically been computed for whole study areas. Fotheringham's study of air passenger traffic in the United States demonstrates how misleading that can be from a geographic standpoint. Coefficients can reasonably be expected to vary over space, much like coefficients of spatial autocorrelation, so why not exploit

the fact? The procedure is analogous to Tobler's proposal (1965) for computing spatially moving means: The regression equation is estimated for observations within a particular distance of each observation taken in turn. Hitherto this technique has not been applied to the sorts of relationship that might be of interest to the human geography mainstream, but it is not difficult to see the possibilities and where they might lead. Furthermore, it is less the technique that is significant and more its recognition that relationships vary over space, suggesting, among other things, areas for more intensive forms of research.

3. *Multilevel models.* In Chapter 5 I talked at length about the question of scale. Again, in spatial–quantitative analysis, there is an interesting history. As I pointed there, the tendency was to allocate variation to particular scales: Total variation could be divided into so much percent at the local level and so much percent at the district level and so on (Moellering and Tobler, 1972). Studies over time (e.g., those of voting behavior across counties and states in the United States) might then provide insight into trends regarding which spatial processes—local level, regional level— were become more or less important. As in the case of spatial autocorrelation and regression, however, the tendency has been to examine what is happening more locally: the variation in regression coefficients across neighborhoods and across regions, for example.

Jones (1991; Jones, Johnston, and Pattie, 1992) has termed the resultant models "multilevel" as a result of the nested, hierarchical form of their data bases. Much as in Casetti's expansion method, the technique involves "expanding" regression coefficients computed for the whole study area as a function of neighborhood, region, and any other level to which the data can be accommodated. This then allows identification of neighborhood- or region-specific effects in the data, much as in the case of geographically weighted regression, but now in a more rigid form in which the "floating" aspect is governed by the spatially hierarchical way in which the data are organized. Jones et al. (1992) demonstrated how this works when applied to voting behavior in England, Wales, and Scotland. So while mining regions have historically been associated with high levels of voting for the Labour Party, the intercept for the relation varies and in one case the direction of the slope is the opposite of what would be expected.

*   *   *

The significance of these developments is, from the standpoint of this book, threefold. First, and as Fotheringham would surely aver, they represent a move away from the sort of generalizing projects typically associated with spatial–quantitative geography, a move away, therefore, from Rushton's "spatial behavior" to what he called "behavior in space." Second, they provide one more set of examples of the spatialization of statistics. Both the expansion method and multilevel models are spatial applications of the statistician's dummy variables and interaction effects. Third, they provide a strong descriptive base for research of the intensive kind, as described by Sayer (Chapter 6, *supra*) and, because of the emphasis on spatial variability, in a more useful form than he might have anticipated.

The Question of Comparative Studies

Comparative study has a long history in the social sciences. In some cases, and most dangerously, of course, the comparative element has remained implicit: a history, an economic survey of a particular country, or an assessment of its political development where the standards of evaluation are being imported from elsewhere and without acknowledgment. Typically where it is a question of some country in the global South, interpretation is susceptible to a Eurocentricity of which the observer is, and in some social sciences more than others, like economics, blithely unaware.

In still other instances, comparison has been the announced objective of study. Stimulated by the financial aid of a federal government keen on waging the Cold War in the 1950s, the academic fields of American sociology and political science in particular had massive interest in what was called "modernization."[12] Modernization was assumed as something that would necessarily unfold within a country, as most notably in W. W. Rostow's *The Stages of Economic Growth*, but it was also recognized that particular social legacies, again usually internal to a country, might make a difference. Evaluating their role would be an important part of the exercise.

Some of this comparative work was quantitative in character. Data were collected for large numbers of countries around the world, and indices were computed and then drawn on in attempt to arrive at conclusions about, for example, democracy or the welfare state (Cutright, 1963, 1965; Cutright and Wiley, 1969)—always of interest to state sponsors—or economic development.[13] Some of this quantitative comparative work focused on what was happening *within* countries. This was particularly the case with political sociologists for whom election results organized geographically provided a ready point of departure. This was what Dogan and Rokkan (1969) would call, after the title of a book that they edited, "quantitative ecological analysis." By "ecological" they meant studies of areal association, and this was exactly the case in the quantitative comparative studies of democracy and economic development referred to previously. Geographical units, countries, or electoral units, therefore, were to be lifted out of their geographic contexts and treated in isolation from one another. The assumption of independent observations that was at least implicit in the regressions of "quantitative ecological correlation" was, in effect, accepted with grim earnestness. Absolute space trumped relative space, which was an intriguing development given that the spatial–quantitative revolution in geography was, at almost the same time, making the shift to more relative conceptions.

There was then a modest countermovement. All of a sudden processes that breached the bounds of the nation state were discovered. The equating of society to the nation that had been taken for granted in the social sciences was challenged. Some of the notable names here were Gunder Frank, Immanuel Wallerstein, and Anthony Giddens. Frank led the way in the early 1970s with his critique of the modernization theory, on which so much comparative study had been based (Frank, 1972). Instead, he reintroduced those metropole–satellite relations that comparative study, in a postcolonial era also marked by the relatively self-sufficient economies of fordism, thought it could ignore. He was then followed by Wallerstein and world systems theory; society, Wallerstein claimed, was defined by the market and the market was worldwide. States entered into a relation with one another in order to defend or challenge positions in a global division of labor. Giddens's theory of structuration then made its mark, processes of time–space distanciation knowing seemingly no bounds and certainly not those of countries.

What is interesting, though, is the fact that despite these interventions, despite the globalization babble from the 1980s on, comparative study continues to enjoy a powerful cachet.[14] Charles Tilly's (1984) influential *Big Structures, Large Processes, Huge Comparisons*, although focusing on theories of social change, does so through a review of different approaches to comparison, only one of which can fairly be said to see countries in relation to one another rather than as isolated or to use the parlance of statistics "independent" observations.[15] Equally influential in the social sciences was the "how-to" manual by Charles Ragin.[16]

For a human geographer, this approach now seems not only empirically questionable, given the nature of the world as we experience it, but also difficult to defend theoretically. One of the major contributions of human geography to the social sciences over the last 30 or so years has been the development of a conception of space as relational: how people and other agents in particular places, whether a city or what we call a "country," internalize through their social relations other places; how, that is, a place, in terms of its institutions, its position in wider divisions of labor, its class relations, its discourse, is composed by wider relations that stretch out beyond what we take to be its boundaries.[17] This conception is very evident in the work of Massey and Harvey, among others, even though they have not used it to critique the concept of comparative studies. Among other things, this is a conception that puts all theories of exceptionalism in doubt. Everywhere is "exceptional," though that does not imply that they might not be different expressions of some underlying ordering process or struggle.[18]

"Countries" are formed from what will eventually become "the inside" as well as "the outside," and it is not always easy to order the networks of relations in that simple binary form; in fact, it is thoroughly inadvisable. Better to conceive "countries" as condensing, like Harvey's "permanences" out of a network of relations that, at the start at least, show no clear geographic structuring. And that "condensation" also raises questions, of course, where "countries" are always for some and not for all and therefore objects of struggle from the very start.

Countries *are* different. Comparison is useful, even essential. It is important to know how countries differ and what some of the correlates of that difference might be. Those differences, though, need to be understood in terms of sets of influences, structures of sociospatial relations that can in no way be reduced to the jurisdictional space corresponding to a country, as having emerged or been reworked within the context of conditions and influences of highly variable geographic scope, or more accurately geohistorical, scope. One result of this is that every country is *sui generis*, and this imposes serious limits on the utility of comparative study. Meanings are going to vary. They may vary within some tolerable set of limits so that, as I said, comparison can tell us something. But vary they nevertheless will. Countries need to be interrogated on their own terms, much along the lines of Sayer's "intensive methods" if causal structures—typically cutting across the boundaries of nation states—are to be identified and more adequate explanations arrived at.

CONCLUDING COMMENTS

Space relations are a necessary aspect of the social process. Social relations are necessarily always spatial, and space makes a fundamental difference not only in the magnitude of their expression but in their very constitution. Despite prognostications of the end of geography or the recent claim that "the world is flat" (Friedman, 2007) space, as in "spacing" or "accessing," remains important. And to the extent that, for example, the real costs of movement go down, then geographical differences hitherto unexploited can assume significance, which is, of course, the story of the geographic division of labor.

Accordingly and inevitably, geography enters into the objects of study and even sometimes into the reasoning of the other social sciences. Yet it enters in imperfectly. Social relations get decontextualized as what applies to one country is assumed to apply to another. "Economies," "states," "societies" get fetishized and their geographic constitution sidelined. Just as human geography cannot function without the other social sciences, which is the basis of the claim that it is "falling apart at the seams" (Johnston, 2005), so does the converse apply. Human geography fulfills an essential role in the academic division of labor: one window, therefore, but an essential window from which to view the social process. It has its own distinctive imagination, sensibilities, and priorities, relentlessly pursuing the spatialization of social theory, of statistics, or whatever else figures in social inquiry. It may indeed be true, as Dicken (2004) has argued with respect to the globalization literature, that the other social sciences sometimes come to dominate a particular area that cries out for the subtleties of the spatial imagination.[19] But the treatment can rarely be adequate since the subject is inevitably approached from such a limited background of geographic learning, experience, and reflection.[20] The pitfalls are manifold, and they will invariably be encountered and ignored. The same applies to the more explicit forays onto the geographer's turf as in "the new economic geography."

On the other hand, in its relations with the other social sciences, one should not exaggerate human geography's uniqueness. If economics has historically been indifferent to space, so too has it been indifferent with respect to questions of power, culture, class, stratification, and all the other things that have become the focus of the other social sciences. And it works the other way around, too; each of the social sciences has created itself as a relatively self-referential whole. Marginal utility theory is foreign to political scientists, sociologists, and anthropologists, even if they have heard of it to begin with. Similarly, role conflict is a mystery to economists. Each of the disciplines, it seems, has grown up around a pluralistic understanding of the social in toto. Given that there was no overriding, ordering theory with respect to the social, each moment of the social process—the economic, the political, the cultural, the social, and indeed the spatial—could be taken as the center of social understanding, isolated and treated as a self-sufficient order; and from the standpoint of disciplinary practitioners developing an identity that they could be happy about, establishing at least an equality if not a superiority with respect to their social science peers, it had to be that way. Any appropriation of ideas from other social sciences has to be on their terms, as in the case of economics and geography briefly reviewed above.

More promising is work that is at least relatively *trans*disciplinary in its intent. Indeed, one might argue that it is not so much the other social sciences that human geography has drawn on, but rather a more transdisciplinary social theory. This clearly started with Marx in the 1970s. It then proceeded through some maverick sociologists, particularly Anthony Giddens and Immanuel Wallerstein; through feminist theory, again transdisciplinary in the form of people like Judith Butler, Donna Haraway, Iris Marion Young, and Nancy Fraser; to theorists of the postmodern, the poststructural, and the postcolonial: thinkers like Foucault, Said, Spivak, and Deleuze, who lack a clear disciplinary home. And to the extent that people with disciplinary identities have been drawn on, it has been those with weaker ones: people like sociologist Michael Mann, anthropologists Jean and John Comaroff, and political scientists like James C. Scott, Michael Peter Smith, and Timothy Mitchell. These, of course, are exactly the sort of people who are less entrenched in a disciplinary culture and so are often willing and able to build bridges of a transdisciplinary sort, as indeed they have to some degree and certainly more than in the case of interdisciplinary connections. Accordingly, there are human geographers whose work is now well recognized outside the field. The names will be the familiar ones. It is this that gives cause for hope for the future—hope that some genuine academic division of labor in which the unity of human experience, simultaneously social, political, ecological, economic, and spatial, will be the starting point rather than something that has been lost sight of. In short, human geography should forget about the interdisciplinary and focus on the transdisciplinary.

Fifty years ago one could not have said this with such confidence. Human geography was isolated from the other social sciences to a degree that was *sui generis*. It had no explicit relation with theory, and it had no recognition of its own wider academic significance, except to complain when historians or political scientists failed a cartography test. Since then, first as a result of the spatial–quantitative revolution and then the socialization of geography, it has been transformed. That in turn raises the problem of the different ways in which we might understand that transformation. It is to that question that I turn in the final chapter of this book.

NOTES

1. Dartmouth to be precise. But there is no Department of Geography at Brown, Columbia, Cornell, Harvard, the University of Pennsylvania, or Yale.
2. The most notorious of these was an attack on geography as a university subject in 1958 in the British *Universities Quarterly*. The reasons would be familiar from those that led to the closure of the Harvard department: "At root, the case against university geography in its present guise is that it has no apparent distinctive purpose and that its claim to a 'holistic' view of reality is no more than an academic masquerade" (David, 1958, p. 273).
3. I use the term "human sciences" deliberately here so as not to exclude more humanist understandings of human geography, in other words, to emphasize its connections not just with the social sciences but also with the humanities.
4. Thomas Friedman's *The World Is Flat* springs readily to mind.
5. "Similarly, many countries' institutions were shaped during colonization, so that examining colonies is a natural experiment. European colonialists found different disease environments around the globe. In colonies with inhospitable germs and climates, the colonial powers established extractive institutions, so that a few colonialists could exploit natural resources. In

colonies with hospitable climates and germs, colonial powers established settler institutions. According to this view, the institutional structures created by the colonialists in response to the environment endure even with the end of colonialism. Thus, the institution view argues that the major impact of the environment on economic development runs through its long-lasting impact on institutions. Technology in this story is endogenous to the institutions that make adoption of better techniques of production likely" (Easterly and Levine, 2003, p. 6).

6. For an example from political geography and an exquisite critique, see Juliet Fall (2010).
7. As in fact was the case of South Africa in the late 19th century subsequent to the rise of gold mining around Johannesburg.
8. For example: Martin and Sunley (1996); Martin (1999); A. J. Scott (2004).
9. Compare Martin (1999): "While spatial agglomeration is a key feature of the contemporary economic landscape, so is structural change. The pattern of uneven regional development is not a static one, but is continually evolving, entailing major qualitative as well as quantitative change. Spatial development patterns do get 'locked-in,' but not indefinitely: the past is always present, but is not all-determining" (p. 76).
10. The complete quotation is as follows: "The essential feature of a useful postulate is that it should describe the rules by which alternative locations are evaluated and choices consequently made. This procedure we may call spatial behavior, reserving the term "behavior in space" for the description of the actual spatial choices made in a particular system. Since behavior in space is in part determined by the particular spatial system in which it has been observed, it is not admissible as a behavioral postulate in any theory. In short, such behavior is not independent of the particular system in which it has been studied. On the other hand, a postulate that describes the rules of spatial behavior is capable of generating a variety of behavior patterns in space as the system of central places, to which the rules are applied, is allowed to change" (Rushton, 1969, p. 392).
11. Though Gould (1970) recognized early on that the fact of spatial autocorrelation was something not to be lamented but to be taken advantage of.
12. An important series was published by Princeton University Press in the 1960s under the heading Studies in Political Development, including among others books on education (James S. Coleman), communications (Lucian Pye), and bureaucracy (Joseph LaPalombara) along with a number of collections. A more programmatic collection was edited by Richard Merritt and Stein Rokkan (1966) entitled *Comparing Nations*. One thinks also of Almond and Verba's (1963) highly influential *The Civic Culture*.
13. A major condition for this was a number of cross-national data compilations put together in the late 1950s to early 1960s. Notable contributions included Banks and Textor (1963) and Russett (1964). Moser and Scott's (1961) *British Towns* was of the same genre but applied to variation within a country and engaged in extensive study of that variation, including classifications of towns. Some compilations of a more conceptual sort, like Merritt and Rokkan's *Comparing Nations* then complemented these efforts.
14. What is called "comparative politics" is one of the major divisions through which American political science conceives itself.
15. This is his procedure of "encompassing comparison."

16. *The Comparative Method* (1987).
17. The work of Coleman (2007) on the ambiguous location of national boundaries as a result of, for example, the mobilization of local governments in policing undocumented immigration adds to the skepticism. Allen et al.'s (1998) *Rethinking the Region* also contains important lessons for how countries might be conceived, as in the possibility of *Rethinking the Nation*.
18. See Callinicos (1988) for a discussion of how this affects the understanding of "exceptional" state structures.
19. See, for example, Sparke's (2009) highly critical review of Saskia Sassen's (2008) *Territory, Authority, Rights: From Medieval to Global Assemblages*.
20. This is, unfortunately, a difficult message to get across. Books by political scientists tend to be reviewed by political scientists, and the same applies to the work the economists and the sociologists. The fact that human geography is a relatively small discipline does not help. On the other hand, the impact in the social sciences in general of two books by David Harvey that have to do with globalization, *The New Imperialism* (2003) and *A Brief History of Neoliberalism* (2005b), suggest that all is far from lost. Likewise, it does not mean that the "findings" of articles in the journals of the other social sciences cannot be challenged on their own ground. As Juliet Fall (2010, p. 146)) has argued, "Geographers need to get beyond the vague feeling of injustice or puzzlement in not being cited and recognized as contributors [to particular literatures—KRC] . . . and engage with those from outside the discipline who think and write about space in such ways. Leaving such claims unchallenged would be to waste a valuable opportunity."

CHAPTER 9

Making Sense of Human Geography, Past and Present

CONTEXT

Practitioners in academic disciplines invariably show an interest in respective disciplinary histories, and geography has been no different. We situate our research with respect to what others have done, though with variable historical depth. We may draw upon that history to justify what we are doing: a contribution to an august theme in the discipline. Or, of course, we may draw upon it in order to justify ourselves by contrast: history as a demonstration of shortcomings that we, on the cusp, are able to avoid, which in turn suggests that writing histories of the field is far from straightforward. In any case, a knowledge of that history, however slanted it might be, is of major significance for disciplinary identities.

Like so much else in the history of the field, the spatial–quantitative revolution is a major divide. Before that, histories had a grimly narrative form; to paraphrase Arnold Toynbee, they amounted to little more than "one damn thing after another." Only with the spatial–quantitative revolution does one start to see more structured approaches to the problem. Initially, this took the form of an application, in fact a *mis*application, of Kuhn's (1962) highly influential notions about scientific change. Later, and setting aside the self-justifying way in which the spatial–quantitative geographers tried to mobilize Kuhn's idea of the paradigm on their behalf, interpretations of changes in geographic thought as a succession of distinct ensembles of theory and method engaged in for their own sake

223

would come to be viewed very skeptically. By then new understandings were emerging, ones informed by the introduction of more critical forms of social theory into the human geography of the 1970s and the realization that power was everywhere and not least in the academy.

It is these new understandings that form the core of this chapter. I build my argument around four central claims. First, all intellectual activity is social; there is a sense in which disciplines are communities of shared but contested understandings, but emphatically they should not be seen as cut off from the wider social world as in the metaphor of the ivory tower. Second, the political is an essential moment of the social; it permeates research, teaching, and publishing within disciplines in the relation between disciplines and to the wider world. Third, geographic thought should be conceived geohistorically: To paraphrase Marx, geographers make their own history but not under circumstances of their own choosing. Academic geography is conditioned by a wider flux of relations, spatiotemporal in character, some of them thoroughly happenstance so that, as per Massey, the concrete trajectory of the discipline can in no way be anticipated. No one expected the spatial–quantitative revolution and no one saw the "posts" and the cultural turn coming. Fourth, while human geography changes, there is also a strong element of continuity in the field; concepts get reworked and meanings change, but there is a residue from the past that is transmitted forward. Furthermore, what looks new, what is presented as new, is often anything but.

In all of these regards, human geography has been no different from the other social sciences. This, however, is little comfort given the way in which they help to shed doubt on the concept of truth. The chapter, therefore, concludes with a discussion of progress in the field: in short, and after what has seemed in the burden of this book as a widening of our comprehension of how human geographies are made, is it actually possible?

SOCIAL STRUCTURES OF GEOGRAPHIC THOUGHT

The notion that disciplinary thinking is socially structured was a major feature of Kuhn's original (1962) claims. His argument was that at any one time a field of inquiry was dominated by a particular ensemble of concepts, theories, and methods: an ensemble that he termed "a paradigm." These concepts, theories, and methods were shared by the practitioners of the discipline and framed subsequent research. New generations were inducted into them through the teaching process. This seems reasonable enough. Students are taught what to see, what to look for in a map or a landscape, what to ask questions about, and how to carry on research. Central to the disciplinary enterprise is, indeed, a process of communication and mutual understanding. This communication structure obviously extends beyond the classroom, though, to include academic journals, publishers, and other arenas, notably conferences. In short, there is much more to social structuring than Kuhn might have imagined.

Not least, there is academic geography's own division of labor. This includes a division within the research process itself between theorists and those specializing in empirical research. There are also the divisions corresponding to the different subdisciplines of political, cultural, social, economic geography, and political ecology. One should also note auxiliary labor. Of primary significance is the publishing industry, which represents some very ambivalent relations with the research process, not least a hard commercial drive that can squeeze out important work, though university presses take up some of that slack. The publishing industry has been noted as an important player (Barnett and Low, 1996), but it remains a research lacuna. The same goes for relations between universities and schools. In Britain, at least, these were historically very close. This worked through a national examination system in which university faculty often played an important role, facilitating the transition from school to university as well as imposing a particular vision of the field from a fairly early age.[1] Finally, under the heading of auxiliary labor, there are the professional organizations. These are important for providing an arena for research and debate in the form of annual conferences and journals. They are also gatekeepers and, therefore, influence what gets published and can, therefore, play a central role in what comes to dominate or what gets to be marginalized.

Where I want to put most of my emphasis in this section, though, is the plurality of social structures involved, something signaled in the section heading. This is because there is more than one social structure to be taken into account. In fact, there are at least three in addition to Kuhn's idea of a community of scientists. To start with, there is the social structure of the academy itself: a structure of disciplines, a pecking order, that has played such an important role in the development of the modern field of human geography. The academy is divided, and power questions aside—issues to be taken up in the next section—this is a problem for any field. As Erica Schoenberger (2001) has argued, every discipline has developed its own culture, its own barriers to interdisciplinary work and to the fruitful cross-fertilization, which is so much touted but so rarely carried through. Among other things, and apart from specialized jargons to which only the trained individual (e.g., geographer, sociologist, economist) is privy, simple matters of language can get in the way. The same word can mean different things. Utility has a very precise meaning in economics that is absent in the other human sciences. Something like role or status in sociology carries its own burden of associations. There are also very different epistemological convictions.

Second is a wider institutional context. This extends both upstream and downstream, though they are often hard to pry apart. The position of geography in the schools has been a well-rehearsed issue. It has been very different in the United States from other English-speaking countries, and this has important implications for the balance at the university level between teaching and research and the possibility of cross-subsidies. Johnston and Sidaway (2004, p. 386) note that at the end of the 20th century British universities allowed for the entry of about 7,000 studying for a degree in geography. In the American case, the closest equivalent to a degree in geography is a major in the field, of whom there were, at their time of writing, only 4,000: a figure of staggering paucity when set beside not just the figures for a country with just one fifth of the population but the membership of the Association of American Geographers, the major organization representing the geographic profession in the United States. In 2010 this stood at over 7,000. In the United Kingdom and the successor states of the British Empire, geography is an accepted and distinct field in primary and secondary education. In the United States this is not the case. This has further ramifications since it means that in the United States the market for geography teachers is exceedingly limited, which means in turn that fewer students want to major in the field. Recently, the rise of geographic information systems and careers in the field has boosted the possibilities of attracting undergraduates, and this has been regarded as manna from heaven for departments that constantly struggle to attain respectable numbers of undergraduate majors much less attract into their classes undergraduates who are *not* majors.[2]

Historically, and for some period after World War II, academic geography in the United Kingdom, at least, seems to have benefited from the disciplinary structure of the university. This was one in which, as Johnston and Sidaway (2004) point out, the competition offered geography from some of the major social sciences, particularly political science and sociology, was quite limited. In the United States these were established fields. In the United Kingdom, particularly in the prestigious universities of Oxford and Cambridge, this was far from the case, which meant that the talent pool that geography could draw on was significantly widened.[3] In the immediate postwar period, this effect was amplified by very generous state support for the secondary and university education of children, particularly those from less affluent backgrounds. This support looms large in the biographies of a number of the British geographers who came of academic age in the 1960s and went on to leave a lasting impression on the field: an untold story.

Third, there is the obvious fact that the social relations among professional geographers are in no way coincidental with all those of which they are a part. They have a life outside the academy. As such, and excepting a certain selectivity of what they are attuned to, especially of a social sort, they are exposed to everything that everyone else is: the media above all but also ways of thinking imparted by parents, friends, and schoolteachers. In this way various understandings of the world can get absorbed in a very taken-for-granted, unexamined way. This is not to say that the media are not an important source of information for the geographer, something that contributes to the archives drawn on in a research project. They may also be where one hears for the first time about some new development in the world: for example, fracking, the competition of smaller towns in Europe for the blessings bestowed by low-cost airlines, or the growth of an expatriate Anglo community in some Spanish Mediterranean resorts. They may also be a mirror of common discursive practice that can serve as an object of interest in its own right: the emergence of new regional concepts like the Cold Belt, regional identities like the French "francilien,"[4] or the standard, almost clichéd ways of interpretation common in the media, as in "environment" versus "employment." Over and above that, though, and much more dangerous in its implications is the way in which popular discourse, whether of the media or of public figures, insidiously communicates a very particular and fragmented view of the world— of things like geography or power or the economic existing as separate spheres, as things in themselves that interact with one another in order to shed light on a particular issue. This, of course, feeds very directly into the highly pluralized approaches to the world that are standard in the human sciences, and on which I have made copious comment in this book.

Emphasis on the social structuring of a field of inquiry in these different ways might be taken as implying a critique of an alternative view privileging the contributions of individual geographers. This would be unjustified. From some standpoints, the role of individuals is very clear. Seminal texts like Törsten Hägerstrand's (1957) *Migration and Area,* Michael Watts's (1983c) *Silent Violence,* Judy Carney's (2002) *Black Rice,* Derek Gregory's (1978) *Ideology, Science and Human Geography,* Doreen Massey's (1984) *Spatial Divisions of Labor,* or Brian Berry's (1967) *The Geography of Market Centers and Retail Distribution* spring readily to mind. There is also the fact of the very uneven distribution of citations to people's work, suggesting that some are far more influential than the vast majority: a log-normal distribution with a very long tail. One should also note a pattern of some, in hindsight major, contribution, which is then taken up and applied by others to new areas. Harvey's influence is very clear in the early work of Watts in political ecology and then later in the reformulation of industrial location and regional development by Storper and Walker. A snowballing of interest can occur, kept alive by further developments, perhaps on the part of the original source of a particular research impetus. The development of Massey's thinking on the significance of what she would later call the happenstance, starting with her early work on the overlay of different spatial divisions of labor, through the recognition of the importance of unplanned locations, to her more recent work on random juxtapositions, is a classic instance of this.

On the other hand, it is equally clear that the attention accorded particular pieces of work owes at least something to their timeliness. They may pick up on some crucial change in concrete geographies, as in Massey's work on spatial divisions or labor, or respond to some contradiction between the geographic research mainstream and problems in the world, which should have been attracting the geographer's attention but were not: Harvey's *Social Justice and the City* is the obvious example, nicely capturing the mood of the time through its title alone. Alternatively, it may be what seems to be a solution to particularly vexing theoretical issues in the field; the roles played by Gregory (1981) and Thrift (1983) as interlocutors of various solutions to the structure–agency problem are cases in point.

Further corroboration for the significance of timeliness is provided by those cases, and they certainly exist, where something is "before its time" or never found a "time" but remains of striking interest—just difficult to relate to what other people are doing or thinking. Harold Mayer (1964) was writing about locational conflict before Julian Wolpert took the idea and situated it in a critique of location theory. Spencer and Horvath (1963) proposed a theory of the production of agricultural regions decades before references to the production of space became the norm in human geography. Particularly interesting is Massey's work on chance juxtaposition. There are clear antecedents, notably Curry's work (1964, 1966) on the random space economy and Hägerstrand's on innovation diffusion (1965) and migration (1957). Hägerstrand even expressed this in more abstract terms in a way that anticipates Massey by at least a decade:

> When members of other disciplines, or laymen for that matter, talk about the "geography" of some area, they probably more often think of the mass of unassorted phenomena to be found there than of the orderly arrangements of selected sets which geographers recently have done so much to try to discern. This latter activity has been extremely productive but perhaps we ought to widen the repertoire to include the study of principles at work when unlike and *per se* unrelated things come together in a mix. (1973, p. 71)

Massey, however, arrived at it by a different route, which almost certainly did not take her through the work of spatial–quantitative geographers like Curry and Hägerstrand.[5] Rather, she was much more attuned to geography's changing intellectual context and its embrace of poststructuralism. This allowed her to specify the idea of what she would call the happenstance in terms appropriate to a human geography infatuated with the idea of discourse: places as the product of the coming together of diverse stories or narratives. Curry and Hägerstrand, on the other hand, were swimming against the tide of a very different infatuation: that of empirical regularities and generalization across locations.

POWER EVERYWHERE

A notable omission from the prior discussion has been the question of power. This reflects Kuhn's view of the matter: There was no struggle; rather, if anything ruled his academic communities it was the explanatory appeal of ideas, in particular the ability of the new to resolve questions that had been raised in the application of the old. But he was very clearly quite wrong, for disciplinary social structures are riddled with struggles to impose particular viewpoints. While academics are certainly driven by a strong sense of the prospect of discovery, there are other motivations that have to do with recognition and material reward. Academic fields manifest very uneven distributions of advantage, embracing among other things the status variations defined by prestigious positions within the institutional structure of the field; the academic hierarchy; positions within professional associations, editorships, and editorial board memberships of journals; but also recognition outside of institutional structures, notably in terms of citations, invitations to give keynote addresses, visiting appointments elsewhere, and even honorary degrees for those at the very peak of the discipline. Nor is it simply a matter of what might be defined as differences of a material sort. In the defense of older viewpoints, in particular, identities are at stake. Who one is, is bound up with intellectual investments in particular sets of theories and methods. There is an emotional attachment to them that cannot be ignored; one is what one does and what one has done for a considerable period of time. In short, whatever way one looks at it, disciplines, their institutions, and their dominant theories and *modi operandi* work at any given time more for some than for others.

On the other hand, disciplinary power structures can change and have changed. The dominance of some can be challenged so that their position in the field undergoes quite rapid eclipse. As a result, they can be replaced by those representing new ideas or new institutional forms or both. The spatial–quantitative revolution is an obvious case in point. Although its advance was eventually slowed down and even stabilized, for a while spatial–quantitative geographers were the flavor of the day. This was certainly more so in some departments than in others, and in some instances a rapid turnover of personnel could transform the balance of advantage in a quite short period of time. My own department at Ohio State was possibly the most dramatic example of this. But everywhere, there were at least some people who wanted to know more about the new ideas, and this resulted in invitations to speak and then, as the ideas acquired more legitimacy, to serve on national committees and eventually to attain the pinnacles of symbolic power as President of the Association of American Geographers, of the Association of Canadian Geographers, or of the Institute of British Geographers.

Even so, this fragility is a variable. Human geography has been more insecure and lacking in confidence at some points in its history than at others, which has meant that the disciplinary establishment was more open to overthrow. It may also be that there is some national variation. Human geography in the United States has been much more subject to challenge to existing ideas and, therefore, to the practices and social structures that have had those ideas as their raison d'être than would be the case of France and possibly of the United Kingdom too. Drawing on the different receptions to "French theory" in its country of origin as opposed to in Anglo-American geography, Juliet Fall (2005) has painted a picture of a quite conservative French discipline, which owes to some degree to a very different institutional structure. Official lists of required course reading are prescribed at the national level, and research continues to occur largely within programs specified by the Ministry of Research and the Centre National de la Recherche Scientifique. The fact that what is in effect a second dissertation ("habilitation") has to be prepared and defended before graduate students can be advised is another constraint on change from the bottom. She also suggests the importance of values of conformity and a surprising (to me) suspicion of innovation.[6] One indicator of the conservatism of the field is what she describes as the virtual absence of a francophone feminist geography.

I have raised this question of the susceptibility of national schools of geography to change because it sheds a somewhat skeptical light on a popular understanding of disciplinary change in geography that was first advanced by Peter Taylor (1976) in explanation of the spatial–quantitative revolution. His central point was that scientists are not just motivated by a search for truth but are affected by their interactions within knowledge communities. Like all communities, scientific ones, as I pointed out earlier, control resources, exercise pressures toward conformity, construct ideologies through which those pressures can be legitimated, and are occasionally subject to internal conflict. Access to advantageous positions depends on adherence to the knowledge community's norms: its particular standards as to what is acceptable as geography, anthropology, or whatever. This adherence comes through research and, most importantly, publication activity. Submission of material that those in power consider "not geography" risks rejection and personal exile to the margins of the community. As Taylor says, "As in all communities the role of an individualist is, at least initially, a difficult one" (1976, p. 131). This is not to say that these decisions are seen as arbitrary. Rather, their essential rightness or wrongness is referred to an ideology, a set of beliefs into which all members are inducted through the academic process, and which sets forth precisely what the acceptable theories and methods and standards of geographic significance and interest amount to. One form in which these ideologies are expressed is the disciplinary history in which the present is, for example, vindicated by reference to the acts of the great founders of the subject.

But despite this unifying force, scientific communities, like other communities, are subject to internal conflict. Perhaps somewhat paradoxically in terms of the system of control that Taylor outlines, those at the bottom of the hierarchy can change the prestige rankings fairly quickly through overthrowing the existing ideology, which legitimates the decisions of those making the appointments, deciding what articles to publish, deciding who is to make important addresses at conferences, and so on. In Taylor's hands, therefore, Kuhn's paradigms become ideologies through which battles for material advancement within professions can be fought. The rise of quantification in geography represented, for Taylor, just such a challenge and, to a considerable degree, a successful challenge. But, and to return to the relevance of Fall's remarks, it was not a revolution that occurred in French geography. There were quantifiers, notably Roger Brunet and Denise Pumain, but significantly neither is cited in the *Dictionnaire de la Géographie et de l'Espace des Sociétés* (Lévy and Lussault, 2003), which is otherwise an excellent guide to debate in French geography.

Harvey's take on the spatial–quantitative revolution was somewhat different from that of Peter Taylor. He recognized that it was in part about, as he called it, "a rather shabby struggle for power" (1973, p. 124), but instead of confining the politics of the field to relations that were entirely bounded by it, he acknowledged the significance of changes and shifting priorities in society at large, in particular how they were generated by the capitalist nature of that society.[7] The implication of his argument is that without those broader changes, the possibilities of struggle within the field itself would have been quite sharply limited. Rather, there was a wider conjuncture of forces that was, in effect, and perhaps without realizing it, exploited by the rising stratum of quantitative geographers in their challenge to the disciplinary status quo.

In short, in his 1973 essay[8] Harvey wanted to give scientific change, including change in human geography, a materialist base. In his view, it has to be situated with respect to the capital accumulation process, its concrete trajectory, and the struggles that it engenders and that form a crucial moment of it. Harvey's point of departure is a critique of Kuhn and his failure to so situate. This accounts for the emphasis he gives to Kuhn's idea of anomaly: observations that cannot be explained by existing theory, contradictions between different theoretical claims within the same paradigm. For Kuhn it was anomalies that were the prelude to scientific revolutions: new ways of looking at the world that would facilitate an explanation of not just what had been understood by the preexisting normal science but also of the anomalies it had turned up. But according to Harvey, Kuhn's treatment begs three important questions: How do anomalies arise? How do they generate crises? How do new paradigms get accepted? The missing term is the relation of science to social need: that anomalies only become significant to the extent that their resolution promises an enhancement of "society's" ability to control and manipulate nature; and that solutions to them then only become accepted to the extent that they facilitate that ability. But as he continues, manipulation and control of nature are not value free: We have to ask, "control for whom"? On behalf of whose interests?

According to Harvey, it is only when one includes consideration of the material base of scientific knowledge that the spatial–quantitative revolution can be fully understood. The "old" based on the qualitative and the unique reviewed in Chapter 1 was not functioning well in terms of the demands coming from the planning field, and a shift to the quantitative and the general promised to be more adequate to that purpose. We can add that from this viewpoint, and as discussed in Chapter 2, the rise of regional science in the late 1950s and the links that quantitative geographers made with the Regional Science Association and its annual conferences and journals was not coincidental. Regional science had a very strong applied bent and generated interest not just among the transportation engineers but also among those concerned with regional economic planning. In other words, what scientists direct their attention toward is not dictated purely by the logic of their theories and methods, as per Kuhn, nor by the struggles for disciplinary power outlined by Taylor, but also by the concrete problems of the world outside: problems of traffic management, urban growth, regionalization for purposes of economic planning, and the like.

GEOHISTORIES OF GEOGRAPHIC THOUGHT

There can only be a situated geography. For geography has meant different things to different people in different places and thus the "nature" of geography is always negotiated. The task of geography's historians, at least in part, is thus to ascertain how and why particular practices and procedures come to be accounted geographically legitimate and hence normative at different moments in time and in different spatial settings.

—LIVINGSTONE (1992, pp. 28–29)

Meanings and the relative importance of concepts have evolved
over time as the power relations forming the field of geographical
knowledge have changed and as new meanings are enforced.
"Power relations" does not refer only to the social structure of
the discipline or the wider society but relates also to the power
exercised by an *epistemic regime* (or system of knowing) which
organizes the whole or part of the field through its dominant
"discursive practices" (favoring of certain concepts and allied
meanings relative to others, association of concepts with others,
modes of writing—narrative versus analysis, topics chosen for
research, etc.).

—AGNEW, LIVINGSTONE, AND ROGERS (1996, p. 9)

The discovery of power in understanding geographic thought was cast to
a very considerable degree with Kuhn as a foil. Central to this were the
arguments made by Harvey and Taylor. Both were intended as very general
statements, which could be applied to any of the more radical changes in
focus and understanding experienced in 19th- and 20th-century human
geography. Both Harvey and Taylor were particularly interested in the
spatial–quantitative revolution and its background conditions, but there
was never any doubt that they intended their arguments to have wider
applicability. Subsequently, and without throwing out the significance of
power relations, these more universal claims have been tempered by an
emphasis on what one might reasonably call the geohistorical or contextual. The works of David Livingstone (1992) and Felix Driver (1992) have
been notable.

Consider now the two quotations at the head of this section. The idea
of power relations emerges to the fore in the second one, and three distinct elements of these are identified: the significance of "the social structure of the discipline" and "the wider society" will be evident from what
has already been discussed in this chapter, though, as Livingstone (1992)
qualifies, "an account that takes seriously the interpenetration of geographical knowledge and broader socio-intellectual circumstances need
not be committed to a reductionist materialism" (p. 27). What is referred
to as "systems of knowing" or an "epistemic regime," though, requires
more attention. At the risk of some slight blurring of meaning here, I am
going to refer to this as a matter of "interpretive framings."

The language—one of discourse, norms, and situatedness, how certain practices come to be accepted as legitimate—clearly has "post" affinities. Accordingly, and as Driver has argued, a contextual approach has
been "more concerned with mapping the lateral associations and social
relations of geographic knowledge than with constructing a vision of
the overall evolution of the modern discipline" (1992, p. 35). Typically, it
involves bringing together all aspects of context identified previously, as

Harvey emphasized in his critique of the paradigm idea, though without his crucial insertion of the role of contradiction, which means that in the Marxist view there is a developmental dynamic and acceptance of foundations to knowledge, which is at odds with the contextual approach outlined by Agnew et al. (1996). I return to this theme later, but for now I want to show how, in fact, this approach might work.

In trying to understand human geography as it was in the first half of the 20th century, we would certainly have to recognize the importance of Darwin and the impact that he had, for good or ill, on thinking not just in the physical sciences but also in the social sciences, and not just in the academic world but also that of the lay world too. Without Darwin it is otherwise hard to make sense of the emphasis placed on the relation between people and their natural environment and on metaphors of adaptation and adjustment; this, we should recall, was the dominant theme in human geography prior to the 1950s. Even when area studies stepped into the breach from the 1920s on, it remained the dominant criterion of significance and the major approach to describing and interpreting the particularity of areas. Regional study would, in consequence, start with the elements of the so-called physical environment before proceeding to those facts of economy, population, and settlement that were supposed to bear some relation to them. At the national scale, the relation between population and resources assumed priority. The most obvious expression of this was the German idea of "Lebensraum" going back to the 19th century, but it was carried forward in the sort of political geography that dominated the first half of the 20th century, all the way from Fairgrieve's *Geography and World Power* (1915) through to Fitzgerald's *The New Europe* (1945).

A second precondition for the geography that dominated at least during the first half of the 20th century was the establishment during the previous century of so-called "geographical societies." These were active in promoting exploration and also engaged in promoting the creation of formal departments of geography in the universities (Unwin, 1992, pp. 79–90). These connections were enduring ones. Even by the 1950s, the *Geographical Journal*, the mouthpiece of the Royal Geographical Society, was still a major outlet for British geographers, featuring the oddest mix of papers that might include, on the one hand, something on changes in the location of population within Great Britain and, on the other, reports of recent expeditions to the Thar Desert of India or to mountain ranges in the Australian Outback.[9] But the connection also helped to further the heterogeneous subject matter that had been at the center of the emerging discipline; exploration was of both the physical and human character of regions hitherto "unexplored."

These particular emphases received further reinforcement from geography's early relation with geology. As Capel (1981) and Unwin (1992) have affirmed, numerous departments of geography were offshoots from departments of geology. This provided the emergent field with a strong anchor in a physical science, further encouraging a focus on the relations between people and nature that resonated with the lay and academic obsession of the time with Darwin and the interest in the relation between organisms and their environment. It also gave further impetus to geography as what has been defined as a peculiarly visual discipline (Driver, 2003; Pocock, 1981). This was further fortified through geographical societies with the interest in exploration and would later see the light of day in the interest first in what was called "field work" and for educational purposes "field studies" and then in landscape. Landscape was a bridging concept that brought together, if rather superficially, both physical and human geography.[10] It also helps account for the fact that as far as physical geography is concerned, geomorphology has historically attracted far more interest than climatology.[11]

THE QUESTION OF NOVELTY

One of the virtues of thinking in terms of geohistorical context is that it brings the present into a relation with the past: Thought is contextualized with respect to a web of relations in space–time, though typically it is time that has been accented, with little reference to how different national schools of geography have contributed to the developing conversation.[12] The point is that there is indeed, and *pace* Kuhn nothing totally new. There is discontinuity in the field but also continuity. Concepts remain recognizable over long periods of time even as their sense relations are changing, either intentionally as they get reworked or unintentionally as a result of other changes in the web of associations. As Agnew et al. argue:

> The [geographical] concepts we have identified . . . constitute key elements of "geographical discourses" that have emerged historically over the past 150 years. There are no original or essential meanings to these concepts that can be privileged over others. Rather, meanings and the relative importance of concepts have evolved over time as the power relations forming the field of geographical knowledge have changed and as new meanings are enforced. (Agnew et al., 1996, p. 9)

New understandings are developed, and particular individuals are important in that development but always on the basis of an existing conceptual corpus. Again, as Agnew et al. have emphasized:

> Key authors have at crucial junctures led discursive innovations that would not have happened or achieved success but for their insight, energy and charisma. One thinks in this connection, for example, of authors such as

Mackinder, Vidal de la Blache, Schaefer, Harvey and Massey. . . . This is not to say that an author is the single "source of significations that fill a work," only that the author is more than a mere "murmur" in a stream of discourse. (p. 9)

It is in this way that the intellectual contexts of disciplines tend to change: not radically in the way that Kuhn believed but gradually, the old being incorporated into the new, the new building on the old and providing it with new interpretations. These new interpretations may be conditioned by changes in the broader interpretive framing, as in the case of the way Darwin was married to geography's long-standing interest in space—not difficult considering the geographic explorations of Darwin and his interest in place specificity, like the unique flora and fauna of the Galapagos.

Specific concepts in human geography have histories. They get developed in accordance with new theoretical developments so as to shed light on changing geographic forms. Allen Scott is well known for his work on the relations between economies of agglomeration, urbanization, and the new landscapes of postfordism. The notion of economies of agglomeration has a history that long predates Scott's work. But what he did was to bring it into a relation with other ideas that were in the air much later. Among other things, he drew on theories of firm governance developed by economists Ronald Coase and Oliver Williamson so as to shed light on the question of the vertical integration/disintegration of firms, a question that he saw as fundamental to the question of why some firms agglomerated and some did not (1985). These arguments would then be drawn on again in an attempt to understand the emergence of what he called the "new industrial spaces" of postfordism (1988).

INTERPRETATIONS

The Spatial–Quantitative Revolution

The spatial–quantitative revolution offers a different set of opportunities for exploring changes in geographic thought. It certainly owed a good deal to the immediate context. As noted earlier, American geography found itself in something of a crisis in the immediate postwar years. Department closures at Harvard and Yale cast a shadow over geography's future as part of the academic division of labor. But there were also new possibilities. Problems in city and regional planning suggested a new future in terms of developing more explicitly the spatial basis of the field and turning it in more analytic directions. Measurement held out the possibility of recasting it in a direction that would allow it to re-present itself as a science. And as Trevor Barnes (1998) has pointed out, the electric calculator made it all the more feasible. What I want to focus on here, though, is the potential for a critical examination of the spatial–quantitative revolution from the standpoint of its discourse and what that discourse said about the power relations in play within the profession itself.

As discussed in Chapter 2, the spatial–quantitative revolution did not go unopposed. There were strong voices of dissent. The battle lines were drawn around a set of mutually reinforcing binaries; ones that, for the most part, were defined by the spatial–quantitative geographers in ways that would work to their advantage. Spatial–quantitative geography was defined as "new," as "scientific," "analytic," "quantitative," "theoretical," and "objective." The "old" geography was, by contrast, "unscientific," "descriptive," "qualitative," "untheoretical," and, through its refusal of the world of quantification and replicability, "subjective." Significantly, what would become the vanguard journal of the revolution was given the name *Geographical Analysis* and carried the interesting subtitle "A Journal of Theoretical Geography." In Great Britain what was initially a yearbook designed to showcase the "new" geography was introduced with the equally telling title *Progress in Geography*. What was happening was self-consciously described as "revolutionary," and Thomas Kuhn's idea of scientific revolutions and paradigmatic change was harnessed to support that view.[13]

These claims were buttressed by a language internal to its practitioners. As Taylor (1976) suggested, part of this was to impress, overawe, and baffle the critics: talk of "inverting matrices," "linear transformations," or testing for "multicollinearity." This was far from the only logic at work. Some of it flowed from the underlying assumptions of spatial–quantitative work. If one believed that the goal was indeed cumulative generalization, then it made perfect sense to talk about "deviant cases" or "residuals." Likewise, the use of regression as a metaphor for cause and effect entailed the use of terms like "independent and dependent variables" or "additive and interactive effects." That this was indeed mysterious to the outsider was true, but it was also a shorthand language for the habitués that had its own coherence.

Nevertheless, the terms used were not entirely innocent ones. The spatial–quantitative geographers tried to put themselves on the side of history. The claims of "science," "objectivity," "theory," and "analysis" were difficult ones to counter, particularly in the technocratic mood of the times. The motivations were diverse. People were undoubtedly taken with the novelty and intellectual provocativeness of what they were embarked on and wanted to share it with others and convert them. They wanted acceptance and they wanted to widen the circle of communicants.

There was also, however, a more conventional political agenda. The demography of the spatial–quantitative revolution is interesting here: it was almost entirely a matter of younger geographers, none of whom were over 40 years of age at the time they became converts.[14] As we saw, Peter

Taylor (1976) argued that in this regard quantitative geography, if not *spatial*-quantitative geography, represented a struggle for power: a way in which those who were not part of the field's establishment and were unwilling to wait so that in the course of time they would be absorbed by it could short-circuit the existing career ladder and push aside the old guard. What was at issue was not just academic acceptance and promotion, important as they were, but placements on editorial boards, elections to office in professional organizations, and the academic appointments of their advisees. Through what they believed to be the revolutionary character of their work, they could successfully scale the ramparts. If that is so, it was a huge bet.[15] At least initially, the establishment, in control of journals and appointments, did not take kindly to these upstarts and their "new" geography. Every spatial–quantitative geographer had his[16] story of manuscript rejection by some journal or other. But beachheads were established in a few departments, and others fell into line in a reluctant desire to have at least some token representation: their own "quantifier." A market for the products of the few departments that were becoming known for their spatial–quantitative work, therefore, opened up. It was then in one of the departments—at Ohio State University—that the new journal *Geographical Analysis* was established with the explicit aim of independence from the editorial decisions of those who, it was believed, neither understood nor wanted to.

Yet despite the attempt to differentiate the spatial–quantitative work from what had preceded it—despite the attempt, therefore, to claim the high ground through references to "science," "revolution," and the "new"—there were important continuities. These would eventually make the binaries engaged with by the spatial–quantitative geographers open to deconstruction. For a start, the spatial–quantitative work shared some aspects of its positivism with what had gone before, suggesting important limitations to its revolutionary character. It talked about "theory," but its theories were based on pretheoretical "facts" just as much as the conclusions of previous generations of human geographers had been. Once its positivism was exposed to critical light, then so would be the self-serving binaries in which it had indulged. The same applies to the way the spatial–quantitative revolution tried to position itself by contrasting its "analytic" qualities with the "descriptive" nature of its "other." "Analytic" carried the sense of "understanding," of "taking apart" so as to expose the underlying "mechanisms." But as would be apparent later, the translation of "quantitative" as "analytic" had its limitations. As Sayer (1984, p. 179) argued, the equal sign in an equation connotes absolutely nothing about causation. Quantitative geography was very good at precise and replicable description, particularly where meanings were unambiguous, but embracing the term "analytical" and so presenting itself as providing the key to causal understanding was a step too far.

The quantitative–qualitative distinction was also something that would be turned against spatial–quantitative geography. During the 1960s "qualitative" had had very definite meanings. It meant "subjective," having to do with the "anecdotal" and the difficult to replicate and most of all, through Hartshorne and the emphasis on the unique "qualities" of particular places, with the singular and therefore what was resistant to explanation. With human geography's increasing engagement with social theory subsequent to the early 1970s, this binary would be reinterpreted. Now the qualitative came to be associated with the causal: with the meanings that conditioned action in the case of humanistic geography, which at the same time subverted the objective–subjective distinction; and in the instance of critical realism, the qualitative became the realm of causal properties and the quantitative the measure of the effects of causal properties when exercised.

A Brief History of Marxist Geography

In anglophone human geography, the interest in Marxism was a major feature of the early 1970s. It was not, however, entirely new. There had been forebears, most notably British geographer Keith Buchanan writing in the 1950s and 1960s. Even earlier had been socialist J. F. Horrabin, who, while not an academic, wrote a book with the title *An Outline of Economic Geography*, which appeared in 1923. He went on to publish an *Atlas of Current Affairs* (1934), which saw several editions (Hepple, 2002).

What happened in the earlier 1970s, though, was arguably different. In part, it was the serious engagement with Marx's writings and then debates in Marxist theory, something that had been previously absent. David Harvey was always to the fore here demonstrating an understanding—indeed a *rapidity* of understanding—of the Marxist categories and logics that was quite *sui generis*.[17] He was not entirely alone in his interest. His students, people like Neil Smith and Dick Walker, would make their own contributions before the decade was out. There were others like Michael Watts and Dick Peet, who showed a sharp appreciation for the way in which Marx could be applied to the sorts of relations, like the ecological in Watts's case, that had long been at the center of concern in human geography. Other names worthy of mention include Jim Blaut whose interests ranged very broadly, economic geographer Mike Webber, cultural geographer Don Mitchell, and Denis Cosgrove, who demonstrated with great imagination the difference that Marxism might make to landscape studies.

It was also the way in which it was part of a broader intellectual landscape in the social sciences running the gamut from economics, through sociology to political science. The almost concurrent emergence of new journals catering to the new radical interest is notable. Geography's contribution was *Antipode*, which appeared for the first time in 1969. For sociology it was the *Insurgent Sociologist* in 1971. In 1968 there had been the *Review of Radical Political Economy*. Political science's new journal was *Politics and Society* (1970). In short, human geography seemed to be capturing something in the air in the human sciences more generally, a radical shift from the technocratic mood that had facilitated the rise of the spatial–quantitative work in the preceding decade. Not all of this was Marxist, to be sure, but at the very least it exhibited strong critical tendencies of a more internal sort. Basic assumptions were not necessarily questioned but distributional consequences evinced a concern hitherto largely absent. This was also evident in human geography's turn to social concern.

How, therefore, might we understand this, what seemed at the time, sudden irruption into a field that had up until then seemed steadily set on the development of what had been hailed as a "scientific geography"? A common view is that it was a reaction to the growing awareness of serious social tensions, partly in the city, as exemplified by the urban riots of the early 1960s in the United States, partly of an imperialist sort and apparent most notably in the highly controversial war in Vietnam. An environmentalist consciousness also seemed to be stirring, pushed along by widely publicized books like Rachel Carson's (1962) *Silent Spring*, Ehrlich's (1968) *Population Bomb*, followed in 1972 by the Club of Rome's report *Limits to Growth* (Meadows, Meadows, Randers, and Behrens). The creation of the Environmental Protection Agency in 1970 gave expression to growing concerns about air and water pollution.

The notion of timeliness requires some qualification. In British geography going back to the 1950s, there had been an interest in regional disparities. The work of people like Peter Hall, Gerald Manners, and Michael Chisholm exemplifies this and was referred to in Chapter 2. This was partly in the context of interventionist policies of the postwar period aimed at mitigating serious unemployment in U.K. areas such as the Northeast, South Wales, and parts of Central Scotland. This then conjoined with the enhanced spatial consciousness that accompanied the spatial–quantitative revolution to provide the basis for the creation of a new journal, *Regional Studies*, in 1966. That this was also an attempt to head off the challenge of Isard's Regional Science Association with its very different, spatial–quantitative agenda should be noted.

Likewise, awareness of an urban question had already started to appear in economics in the 1960s, notably with the emergence of a distinct subdiscipline of urban economics. This certainly had a very orthodox wing, as in the work of people like Raymond Mills and Richard Muth. There were others, though, like Matthew Edel and Jerome Rothenberg, whose way into the dysfunctionalities of the city was largely through the challenges to the orthodoxy coming from recognition of how space disturbed basic assumptions. For many, externalities became a key concept in a more critical urban economics. Part of this came from the publicity given to issues of air and water pollution, to be sure, but there was also intersection with questions of jurisdictional fragmentation in the city and, therefore, with sharp inequalities of life chances. One expression of this was the interest shown toward the end of the 1960s in the so-called "transportation and poverty" question, which in the hands of geographers would become the problem of "spatial mismatch."

This urban economics literature was highly influential in the emergence of radical geography. This reflected the way in which the people who were significant in this came largely from a background in spatial–quantitative geography, where the economic had always been privileged. Harvey's essay on "Social Processes, Spatial form and the Redistribution of Real Income in an Urban System" (1971) is of crucial interest. In his own words, this was an expression of his liberal phase, where he had cast off the clothes of a quantitative geographer but had yet to assume those of Marxism. Part of his critical foil is provided by the urban economic orthodoxy represented by people like William Alonso, James Buchanan, Mills and Muth, but he also draws heavily on the contributions of those of a somewhat more critical bent like Otto Davis, Julius Margolis, Edward Mishan, Mancur Olson, and Harvey Perloff. But Harvey was by no means alone in drawing on this source, as is evident from a quick perusal of some of the volumes that appeared in the McGraw-Hill Problems Series in Geography.

In short, the conditions in terms of real-world political issues and a shift in the center of gravity of the human sciences in a more critical, more radical direction were highly propitious for the emergence of Marxist geography. Harvey's genius not so much in seizing the moment but in his constant and restless search for a critical angle on the world and his ability to read and make sense, including a geographic sense, of Marx in a remarkably short space of time was also a major contributing factor to what happened. There should be no doubt about the significance of his contributions both within and outside geography. He would have his disciples during the 1970s, along with those who followed the more modest "critical mainstream" that I described in Chapter 3. Yet by the turn of the decade and despite the continuing contributions of those who had made their mark earlier, the impetus was beginning to wane. I do not mean that it did not continue to attract attention. At national geographers' conferences, people were always interested to hear the likes of Dick Walker, Neil Smith, and Michael Watts,[18] but I suggest the interest was superficial. The active embrace of historical materialism and its extension to the interpretation of the geographic was, for whatever reason, insufficiently attractive.

So what happened? The obvious answer is a further shift in dominant epistemic regimes in the social sciences heralded by the interest in post-modernism, poststructuralism, and then postcolonialism. This was sharply orthogonal to virtually everything that Marxism and Marxist geography stood for, not least its assertion of a material base for social relations, its claims to universality, and its totalizing character. Yet as a Marxist understanding of the shift, this will not suffice. In particular, we need to ask exactly how the world had changed to make the "posts" attractive points of reference. The answer that Harvey himself provided in *The Condition of Postmodernity* in 1989 was the sensory impression of the fleetingness of the world left by the speedup of a time–space convergence logically entailed by capitalist development.

Harvey was particularly impressed by the disorder and uncertainty entailed by the collapse of fordism and the move into what some called "flexible accumulation." Eagleton's (1995) approach was somewhat different: an emphasis on a widespread sense of the permanence of capital, its seeming invulnerability and oppressiveness, and the other side of the coin: the decline of the labor movement. He couples this with an attempt to understand the appeal of postmodernism to younger generations of dissident academics who grew up, necessarily, with a limited historical memory of the labor movement's achievements, even if qualified ones, and of its one-time vigor, which may well account for the disinterest shown by the "posts" in class as an axis of difference. I add, though, that it is not just the deliquescence of the labor movement that should attract attention but also an uneven development that caught up with it subsequent to the 1960s: in short, that overwhelmingly masculine and white character, which made it vulnerable to what Hobsbawm (1996) derided as identity politics and that would come horribly home to roost with the collapse of those industries in the late 1970s and 1980s, which had been the major support base of the labor movement—the railroads, iron and steel, coal, the docks, shipbuilding—and the concomitant rise of the new right and neoliberalism.

Geographically Uneven Development and Classical Geopolitics

The spatial–quantitative revolution positioned itself with respect to what it liked to see as a predecessor lacking in serious intellectual credentials. In this regard, it would be no different from other shifts in the changing terrain of post-1950s human geography. For the Marxists of the 1970s, the "Other" was spatial–quantitative geography. Rather than "prescientific," the new epithet was "social irrelevance." Instead of *Geographical Analysis: A Journal of Theoretical Geography*, we had *Antipode* with its deliberate message of a hidden world, an underside, and the cover image of the chains of oppression being broken with a sledgehammer (see Chapter 3). And so it would go: the attack on "jumbo Marxism" important to establishing the credentials of society-and-space and the still later charges from those enamored of the postmodern against the "essentialists." Eventually, there would emerge a "new geopolitics" defining itself with respect to its older, imperialist, racist, gendered forebear. And indeed it was all of those.

Interestingly and perhaps significantly, the classical geopolitics that has been in the sights of critical geopolitics and that was associated in particular with Mackinder and Bowman did not seek to justify itself by a rejection of any other approach within human geography. It would not even have recognized itself as an approach. In any case, its audience lay only in part in human geography. To a very considerable degree, it lay elsewhere in the protagonists of what has since become known as state-craft.

Nevertheless, as a discourse, it had clear roots in the dominant assumptions of the age: notably ones of white superiority and of struggle for global hegemony, though with some interesting national variations. For Mackinder, empire was taken for granted. Bowman the American, in common with a supposedly anti-imperial identity, albeit not especially strong, was more fastidious. He could not recognize American imperial-ism when he saw it: either the imperialism of westward expansion within the United States or the trade imperialism that he was to advocate in his 1928 book *The New World*. What I want to emphasize, though, are the material roots of classical geopolitics. Every identifiable school of thought or approach has them. This is apparent in remarks that I have already made about the spatial–quantitative revolution and its relation to urban and regional planning and the technocratic mood of the time. The more recent interest in networks and flat ontologies owes a great deal to notions of flexible accumulation and globalization, which have their own material referents. Classical geopolitics likewise is firmly rooted in the material puzzles of its age. Above all, these were the related dilemmas of class struggle, geographically uneven development, and global hegemony.

In the English-speaking world, the person most closely associated with the birth of geopolitics is British geographer Halford Mackinder. His notion of the Heartland, its implications for world domination, its appro-priation by German geographers and in particular by Karl Haushofer and his *Institut für Geopolitik*, and then its supposed influence over Nazi war plans (Pounds, 1953; Downs, 1956, Chapter 9) almost guaranteed that. Less associated is the name of American geographer Isaiah Bow-man, even though his work in political geography is widely known. This is unfortunate since his geopolitical ideas, while more fugitive in their exis-tence and in contrast to the leaps of imagination of Mackinder, even hum-drum, were indeed there and afford a fascinating counterpoint. The ideas of both men are firmly situated in the shifting contours, as they were at the close of the 19th century and the first two decades of the 20th, of geographically uneven development: in particular the way in which Great Britain was henceforth on the defensive while the United States could look forward to the possibility of global hegemony. Mackinder was to be the voice of empire challenged and Bowman that of empire in waiting.

As David Harvey (1985) has emphasized, the politics of geographically uneven development is a class politics. At the heart of the accumulation process is the class relation. It is the constant pressure of the working class on capitalist profits that ignites competition. This in turn is an essential condition both for geographic expansion in various forms of imperialism and for the struggle between place-specific coalitions of capital. As that competition works its uneven geographic effects, so the capital–labor relation can become increasingly fraught.

In the view of many, and as discussed previously, by the end of the 19th century the possibilities of imperial expansion had been severely curtailed. This is the problem of closed space that Gerry Kearns (1984) has emphasized.[19] This was noted by Mackinder:

> Whether we think of the physical, economic, military, or political interconnection of the Globe we are now for the first time presented with a closed system. The known does not fade any longer through the half-known into the unknown; there is no longer elasticity of political expansion in lands beyond the Pale. Every shock, every disaster or superfluity, is now felt even to the antipodes. . . . Every deed of humanity will henceforth be echoed and re-echoed in like manner round the world. That, in the ultimate analysis, is why every considerable State was bound to be drawn into the recent War, if it lasted, as it did last, long enough. (Mackinder, 1919, p. 40)

This also formed an important aspect of Bowman's thinking. As Kearns pointed out, American historian Frederick Jackson Turner was also strongly taken with the closure, as he saw it, of space. The particular form in which this interested him was less its implications for imperial expansion and more the closure of the American frontier. This was to be formative in Bowman's thought and his advocacy, in his book *The New World*, of a new commercial imperialism, at the head of which would be the United States, and replacing the old territorial one whose future was of such concern to Mackinder.

The class question was a central one. Turner's argument had been that the closing of the American frontier would be a challenge for American democracy. The frontier had acted as a safety valve for class tensions in the eastern, industrial parts of the United States. Movement to the frontier to homestead had, in his view, created a scarcity of labor power. This allowed increasing wages, an increasing standard of living for the American working class, and a subsequent dedication to democratic compromise. All this could now be in question. For Bowman it already *was* in question. We should recall the massive growth of worker unrest in the United States and in the world more generally that culminated in the Bolshevik revolution in 1917 and subsequent, if abortive, revolutions in Germany and Hungary, something that Beverly Silver (2004) has documented statistically. The answer for Bowman, laid out in the final chapter of the revised edition (1928) of *The New World,* was commercial expansion overseas:

> The West will no longer furnish an outlet for eastern population, except indirectly by reason of the increased economic demands it makes upon the East and particularly upon its industries. Eastern social and industrial problems [*sic*] cannot be solved in the historical manner by a flow of population to another region. They will have to be met in the fields of their origin. We conclude that the creative energy, initiative, enterprise, and spirit that have led to the occupation of the land of the United States must now be expended upon a new group of problems which result from denser populations that have no outlet upon cheap land. (p. 690)

And:

> We must come at last to more intensive work upon the land and a lower standard of living, *unless we develop our foreign markets in competition with European nations* long in the field and expert in the use of cheap native labor and the production of crops in every zone. (p. 691; my emphasis)

It is these facts, he argues, that explain the increasing international involvements of the United States (p. 693); and of course it was to facilitate the "increasing involvement," to provide it with a firmer basis in understanding of other countries, that Bowman had written *The New World* in the first place.[20]

This is the voice of confidence—a confidence rooted in the growing share of world production assumed by the United States and by its increased ability to compete with British industrial goods in the rest of the world. For Great Britain, it was otherwise. Instead of confidence there was a sense of narrowing horizons. Embracing the doctrine of free trade had served British purposes well when it had enjoyed advantages of industrial productivity over its major competitors in the world. But that was no longer the case. Rather, it was a matter of *sauver ce qu'on peut.* Instead of the free trade on which the United States would seek to achieve global hegemony, there was a growing call in Great Britain for what was called "tariff reform," and Mackinder was part of that (Semmel, 1960; Tuathail, 1992).

Free trade had become a new bogy. Tariff reform was seen as a means of defending empire. The goal was to protect British industry by placing a tariff on foreign goods entering the empire. This, the reformers believed, would also have important social effects, mitigating class tensions. Employment would be better paid and unemployment would diminish. At the same time, the customs revenues would be used to finance social outlays, including pensions. In a turn of the virtuous circle, a better paid working class would also be a better nourished one, an issue of the quality of imperial manpower that had surged to the fore in the context of the Boer War.

This was a program that Mackinder supported, but through an exploration of the geopolitical implications of free trade, both internationally and domestically, he put an extra gloss on it. On the one hand, free trade meant intensified specialization and this in turn implied a search for foreign markets, heightening the tensions with other imperial rivals.[21] On the other, it intensified class struggle. Nations, he argued, could be organized by classes or by localities. These were the alternatives. What destroyed localities was specialization:

> It is the principle of *laissez-faire* which has played such havoc with our local life. For a hundred years we have bowed down before the Going Concern as though it were an irresistible God. Undoubtedly it is a Reality, but it can be bent to your service if you have a policy inspired by an ideal. (1919, p. 247)

He then goes on to explain:

> By proper control you could have substituted a "village region," with a group dependent on each factory or group of small factories, wherein rich and poor, masters and men, might have been held together in a neighborhood responsible relationship. (1919, p. 249)

For Bowman the closure of the frontier meant intensified class struggle. The solution would be an increasing national orientation toward foreign markets, which would in turn require access to them, and in the course of time, as we now know, the United States would replace Great Britain as *the* advocate of liberalized trade. For Mackinder, closure—this time on a world scale—meant threats to British hegemony. The solution was a withdrawal from free trade into a world of more managed national growth; liberation from the "Going Concern" that was specialization on an international scale and which was fomenting the international organization of the classes. This in turn would allow a solution to working-class challenge specific to the British situation.[22]

This is a far from a comprehensive analysis of the original, classical geopolitics, of course. There were features of the epistemic regime that I have deliberately excluded. As Kearns (1984) has indicated, Mackinder emphasized the relations of people to the natural environment and the national struggle for existence, which had been a development of social

Darwinism. Bowman was much the same.[23] Neither could be accused of environmentalism, though Bowman was perhaps the more judicious of the two.[24] Rather, my purpose has been to bring out the way in which the early geopolitics was conditioned by the material dilemmas posed by a very particular geohistorical context.

It is also worth returning to the particular positionality of this classical geopolitics. As remarked earlier, it is very different in this regard from that of the spatial–quantitative revolution or anything since. It was not trying to prove a point within the closed world of academic geography suggesting, for example, their irrelevance to the world that they lived in. For a start its constituency was always more plural. Both Mackinder and Bowman had wider, more lay, yet politically informed audiences.[25] This meant that the points that they were trying to score were very different ones and were with respect to counterparts *in other countries*. What was at stake was their relative scientificity. In a fascinating article, Bowman (1942) wrote critically and at length about the German school of geopolitics, which he believed had been perverted by subordination to the Nazi cause. Among other things, it involved the use of maps for propaganda purposes and drew on Mackinder's Heartland theory to justify German war strategy. Bowman characterizes "Geopolitik" as "pseudoscience." Yet today, such a critique would be looked at more skeptically. "Science" too has come into critical focus, expressing notions of authority and desires to marginalize alternative understandings of the world. Again, Bowman, un-self-consciously, expresses a particular position: that of a nation at war with Germany and one anxious to justify its participation in that war. Bowman refers to notions of the pacifism of peace-loving nations and how the world has to be made safe for democracy, but as later interpretations have made clear (Mandel, 1986; Smith, 2003), the participation of the United States in World War II was anything but innocent.

THE QUESTION OF PROGRESS

There was no doubt in the minds of its practitioners that the spatial–quantitative revolution represented a leap forward in the history of the field: evidence that progress was possible. Who could doubt that making geography more "scientific," observing standards of replication, an embrace of objectivity in all its senses could not mean that our understandings of the world were improving, and at a rapid pace? One of the ways in which this was argued was, indeed, that a prescientific paradigm had been replaced by a scientific one. But in embracing the idea of the paradigm and by explicit reference to Kuhn, the seeds of doubt were already being sown. This is because for Kuhn there could be progress only *within* a given ensemble of theory and method and not between. What he called "normal science" would prevail under the aegis of a particular paradigm. Knowledge could advance in a cumulative fashion, a predictive capacity could be achieved but eventually anomalies would show up: observations that could not be reconciled with the prevailing theory or inconsistencies in findings. This would be the condition for the development of an entirely new paradigm capable of bringing the anomalies and inconsistencies under its explanatory umbrella, at which point the labor of normal science would start again but without reference to what had been seemingly achieved when working with previous paradigms. This is a crucial part of his argument and stems from Kuhn's assumption of what he called incommensurability. In the extreme, it implies that the protagonists of different paradigms, to the extent that they coexist, would not be able to communicate with each other: There would be no terms that they shared and that might provide points of reference in arbitrating differences of view. His notion of progress in science was, therefore, a highly qualified one.

Skepticism regarding the possibility of progress has been deepened by postmodern and poststructural thinking. Conventional histories of the sciences celebrate the onward march of the modern and bringing the natural and social world within human grasp so as to further human emancipation. Like all grand narratives, this is another case of power–knowledge: The power of some is furthered by the spread and internalization of a particular discursive understanding. There are other discourses but, given the lack of firm foundations for knowledge, no way of arbitrating them. The idea of truth and, therefore, of some sort of progress in knowledge is left in the dust.

One of the few people to develop a concerted retort to these claims has been Keith Bassett (1999).[26] His argument is from a distinctly critical realist position. He emphasizes the way in which observations are conceptually dependent and not conceptually determined. So that we always stand to be surprised by the turn of events and so forced into revising our ideas. This might seem to endorse Kuhn's view of the significance of anomalies to paradigm change; when concept dependence, in effect, trumps concept determination of observations, a wholesale reordering of the theoretical landscape is required to the point at which nothing in the old one is recognizable anymore. The problem of incommensurability is highly exaggerated, though. In order for frameworks of understanding to be in contradiction to one another, and for the concepts of the new to be comprehended, something in the sense relations of objects has to endure. With the advent of spatial–quantitative geography, the concept "accessibility" had to make some sense to those trained in the more qualitative, region-focused human geography that had preceded it, and all converts had that in their background. And, of course, it did. Received ideas of accessibility might not have included such meanings as connectivity and time–distance, but they could be expanded quite easily, particularly given familiarity with the ideas of time and connection.

What is missing from this response to those who would like to dispense with the idea of scientific progress is the idea of contradiction. Ideas about relative space had been around a long time before the spatial–quantitative revolution came along and then placed them at the center of its theories. It is not merely that human geographers had a history of drawing on them, if not in a particularly systematic and concerted fashion. It is also the fact that in an economy structured around exchange over space it is difficult to see how anyone, lay or academic, could have avoided recognizing their significance and developing some abstract understanding of relative space, if only for navigational purposes. But it took an intersecting set of contradictions for human geography to reflect that elementary fact.

Harvey (1973, Chap. 4) rightly emphasized the contradiction between the increasingly problematic nature of relative space and human geography's technical and theoretical inadequacy. Scott (1982) added something to that by drawing attention to the way in which the expansion of the state had intensified popular expectations regarding policies that might draw upon an improved technical and theoretical apparatus. One might add that there were also contradictions within academia that were relevant in the fact that the human geography of the immediate postwar period seemed unable to contribute anything useful to the division of labor in the social sciences and so while remaining *in* the university could hardly be said to have been *of* it. The way in which human geography has been transformed over the last 50 years or so suggests that that is no longer—or at least *should* be no longer—an issue. The imperialism and self-sufficiency of the social sciences remain a problem, but now at least human geography has the intellectual weapons to mount a convincing response and even a counterattack.

NOTES

1. Historically this would have been through the exams for the General Certificate of Education at the Ordinary Level, normally taken at about the age of 15 or 16.
2. Most undergrads taking geography courses in the United States are majoring in other fields and fulfilling what is called "a general education requirement" for their bachelor's degree or taking a course as an elective. Most undergraduate majors are recruited via those routes.
3. This helps to account for the major impact on Anglo-Saxon geography of Cambridge graduates like Peter Haggett, David Harvey, and Michael Webber and Oxford graduates such as Dick Chorley, Doreen Massey, and Allen Scott.

4. Someone residing in the Ile-de-France and so testifying to the emergence of a wider metropolitan region much like "London and the Southeast."

5. Even more stunning is the following quotation: "If everything occurred at the same time there could be no *development*. If everything existed in the same place there could be no *particularity*. Only space makes possible the particular, which then unfolds in time. Only because we are not equally near to everything; only because everything does not rush in upon us at once; only because our world is restricted for every individual, for his people and for mankind as a whole can we in our finiteness endure at all . . . Particularity is the price of our existence . . . The mighty elements of spatial discipline tend to preserve geographical and cultural roots in spite of freedom." It sounds like it just might—but only just, given the reference to distance decay—be Massey, but it is in fact Lösch (1954, p. 508).

6. As she remarks, "This is not a system designed for rapid innovation or the rise of freethinkers—innovation for innovation's sake is scorned upon and pointed out as something uniquely Anglo, and therefore intrinsically suspect" (Fall, 2005).

7. For a similar argument, see Scott (1982a).

8. Chapter 4 of *Social Justice and the City*.

9. Significantly, Peter Haggett's first "quantitative" study (1961) appeared in the *Geographical Journal*.

10. The role of landscape in human geography should be clear. But consider also titles like Dudley Stamp's *Britain's Structure and Scenery* (1946) and Gordon Manley's *Climate and the British Scene* (1952).

11. The earlier affinities with geology are presumably also of significance in this regard.

12. Or perhaps this is an acknowledgment of how weak those relations have been. As far as Anglo-American geographers are concerned, French and German geography, let alone Italian or Spanish, are virtually *terrae incognitae*.

13. There were, as Mair (1986) later pointed out, serious ironies in what was actually a misappropriation of Kuhn's arguments. The idea dear to the heart of the spatial–quantitative geographers that what they were doing represented "progress" did not jibe with Kuhn's notion of paradigms as generating noncommensurable forms of knowledge. His relativism also struck a harsh note alongside the positivism of the spatial–quantitative work.

14. It was also very masculine. One can say that almost everybody and everything in universities was at that time. Yet there are numerous anecdotes suggesting that the attitude to female quantifiers was not especially friendly.

15. It may also have been a bit more complex than Taylor is suggesting. First, the people who led the revolution—like William Garrison, Harold McCarty, and Edward Taaffe—were not the youngest. They were certainly the most likely to lead it because as faculty members they already had their foot in the professional door. But the spatial–quantitative revolution gained rapid and critical mass from the graduate students who, by whatever route, found themselves in the world of spatial–quantitative geography. They were susceptible not necessarily because they wanted a speedy route to the material perquisites of professional life but because their identity with the old geography was considerably weaker than older cohorts. For the latter, accepting the new

doctrines would have been much more difficult: an assault on all that they had learned to hold dear and, therefore, an assault on them as persons.

16. As per note 14, I use the term "his" deliberately.

17. So much so that he would be eventually lauded for his highly original insights in Marxist circles more generally: among other things, a widely sought-after keynote speaker at conferences and winner of the prestigious Isaac Deutscher Prize in 2010. His *Limits to Capital* (1982), though somewhat ignored at the time by nongeographers, is now widely regarded as a major contribution to the Marxist corpus.

18. Though their qualities as public speakers also made a difference.

19. These were ideas that were "in the air" at the time, as in Rosa Luxembourg's pessimistic views about a world closed to further imperial expansion. But, as Kearns points out, they also articulated nicely with the crudities of social Darwinism and, I add, quite explicitly in the case of Ratzel and his equating of state to organism and the organism's need for space.

20. "To face the problems of the day, the men who compose the government of the United States need more than native common sense and the desire to deal fairly with others. They need, above all, to give scholarly consideration to the geographical and historical materials that go into the making of that web of fact, relationship, and tradition that we call foreign policy" (p. iii).

21. "The Cobdenite [i.e., free trader—KRC] believes that international trade is good in itself, and that specialization as between country and country, provided that it arises blindly under the guidance of natural causes, should not be thwarted. The Berliner, on the other hand, has also encouraged economic specialization among the nations, but he operates scientifically, accumulating in his own country those industries which give most, and most highly-skilled employment. The result is the same in each case; a Going Concern of Industry grips the nation and deprives it, as well as other nations, of true independence. The resulting differences accumulate to the point of quarrel and collision" (1919, pp. 230–231).

22. Both visions, those of Bowman and Mackinder, have to be evaluated in terms of differing positions with respect to the changing geography of uneven development. Different again was Germany, and we can gain added perspective on the emergence of geopolitics by brief reference to that case. The important name here is Friedrich Naumann. In his 1915 *Mitteleuropa*, he urged the creation of a politically and economically integrated Central Europe under German rule along with programs of Germanization and German settlement in the Crimea and the Baltic States. The exploitation of this region by German industry would then facilitate competition with the British Empire. This would also, he believed, in an interesting if unintended nod to Bowman, enhance social peace in Germany itself through the combination of an escape valve in the form of new areas of German settlement and through the increasing wages that would flow from improved access to markets for German products.

23. For example: "This is a competitive world and we shall long continue the evolutionary struggle that marked the rise of mankind from the primitive state. National and racial ambitions and rivalries will continue to the end of time, though they may be greatly reduced in scope and intensity" (p. 4).

24. In the British case: "Huge coal deposits and an island base have supplied

two prime geographical advantages, but they have contributed power only through English character" (p. 35). And in France: "To appraise the power of France in modern Europe one must look below the surface of things, at some of the fundamental bases of life in soils and minerals. Above all one must look at the character of the French people, for it is they, rather than the natural resources, that in the last analysis make their land what it is" (p. 160).

25. As I remarked previously, this was explicit in Bowman's *New World*. Mackinder's *Democratic Ideals and Reality* was in part a contribution to the ongoing struggle around tariff reform and in part a reflection on the meaning of World War I for the future of Great Britain.

26. He begins his article with the playful suggestion that if the skeptics are correct "it would seem a heroic but increasingly empty gesture to go on calling a journal *Progress in Human Geography*. Perhaps it is time to rename the journal *Endless Reinterpretations in Human Geography* or, even more simply, *Summaries of What Academics Have Been Doing to Further Their Careers since the Last Issue?*" (1999, p. 28).

Postscript

The transformation of human geography since the late 1950s is little short of remarkable. It recalls Allen Scott's (2000) celebration of "the great half century" when talking about economic geography. Human geography has been converted from an academic backwater that had difficulty representing what it stood for into a vibrant subdiscipline, thriving on vigorous debate, even a contentious field, and far more confident regarding what it is about and its ability to fulfill that promise. It is useful to recall here just how dismal things were some 60 years ago. Theory and method were almost foreign words. What theory there was, was more about determining criteria of significance than for identifying areas for investigation, for providing a source of hypotheses, and as a means of arbitrating competing claims. Method began and ended "in the field." Field study was not something that could be taught. Rather, it was as if it was something to be absorbed through the pores of the skin. In part because of these self-imposed limitations, in part because of the fixation on the relation between people and the so-called "natural environment," human geography found itself separated from the other social sciences in a world of its own. It had nothing to say to them and very little to learn from them and in Great Britain, at least, where it did not undergo the humiliation of expulsion from the more prestigious universities, self-satisfied. It was isolated, self-referential, and for the most part happy to be so.

Sixty years later it is a very different creature. To say that it is unrecognizable would be unfair. There are conceptual affinities. But how space, nature, people are thought about and understood is now very different. Space, and in accord with the atheoretic character of the field

253

was something naïvely given: a matter of the location of settlements at bridging points, of shifting strategic advantage for national governments, of points of ease of assembly for the raw materials used in a particular industry. Today the distinctions between absolute, relative, and relational space have become part of the common argot of the field, and in their turn opened up new ways of viewing old questions like those of boundary effects or scale, as discussed in Chapter 5. There is also a realization now that spatial relations can be the basis of a theorizing of the social in ways whose subtlety could only be dimly apprehended some 40 years ago as human geographers started to talk about the relations between society and space.

What set the transformation in motion was without doubt the spatial–quantitative revolution. As I have tried to emphasize, it was the great watershed of anglophone human geography in the 20th century. The field would never be the same again. It brought theory and method into its common vocabulary and set of concerns for the first time and set off a ferment of discovery. The object of that ferment—spatial organization—might in hindsight seem limiting, but at the time it was approached as not just something that could be useful to the applied sciences of urban and regional planning but as a thing of beauty: a symmetry in the way things were arranged over space, the way in which people organized their lives over space, that was remarkable. Human geographers started looking at the world with very different eyes: Not quite Kuhn's paradigm shift, but one can see why contemporaries might have been seduced into thinking that it was, and quite apart from their own disciplinary ambitions.

For a while it seemed as if the "new" geography would conquer the field and become thoroughly hegemonic. But it was clearly not to be. Through the emphasis on theory and method and a grasp of the social, if rather narrowly conceived, it opened up the possibility of critique: a possibility that would then be realized as the limits of what had so far been achieved became more apparent. This was human geography's embrace of more critical variants of social theory: a development that has gone through numerous stages, starting with Marxist and humanistic geography and then receiving further stimulation through the interest in the various "posts." The field has been revolutionized, opening up further possibilities that lay well beyond the capabilities of spatial–quantitative geography. It has, in short, been an exciting time to be a human geographer.

Even so, the transformation has not been entirely without cost. In this book I have often used the terms "human geography" and "geography" interchangeably. This conflation, of course, obscures the fact that there is something called physical geography as well as the way in which the field of geography at one time tried to bring them together. This was through the tactic of putting people–nature relations at its center. But this had very severe limits. The physical geography that one needed in order to make sense of human geography was very limited. One could acquire some sense of climatic variation without an understanding of air–mass climatology; and one could talk of the significance of river terraces for agriculture without getting into the geomorphology of so-called "underfit" streams. The converse applied. Human beings may indeed have become the most significant of all geomorphic agents in terms of the speed with which they can transform the shape of the earth but, again, you do not need much human geography to take that into account in geomorphological studies. Nevertheless, training in geography 60 years ago was in both human and physical geography, and many professional geographers prided themselves in the breadth of their knowledge. It was not that unusual to find someone giving courses on both the regional geomorphology of the Scottish Highlands and the historical geography of Central Europe. Again, this had its limits and the spatial–quantitative revolution was seen as a way of bringing the two subdisciplines closer together. But despite a few striking attempts, like the book *Network Analysis in Geography*, written by geomorphologist Richard Chorley and human geographer Peter Haggett, its potential once more was found to be very limited. Yet as Doreen Massey (1999b) has tried to emphasize, it is a dialog that could still bear fruit and that, therefore, should be encouraged (Cox, 2006). Like human geography, geomorphology also is a "complex, historical science" with all the attributions I gave that term when applying it to human geography in Chapter 7.[1]

A second problem is the marginalization of quantitative methods in human geography. Given the way in which the original stimulus to change in the field came from that direction, this is, of course, supremely ironic. It should also be cause for regret. Quantitative methods are of major importance in providing a synoptic view on what is out there: the shifting dimensions of something like geographically uneven development or, to take something much more particular, the rural turnaround. Or again they can be used to identify the particular cases to which other methods are to be applied in trying to identify causal structures, as per Sayer's distinction between extensive and intensive forms of analysis discussed in Chapter 6. This means a shift in the direction of what John Tukey (1977) called "exploratory data analysis"; ransacking spatial data for what they can tell one about relationships and deviant cases that might stimulate investigation of a more focused nature. Of course, quantitative methods are limited in what they can tell one, but so too are so-called qualitative methods. They need to be used in tandem. Their power, moreover, is by no means a function of their complexity. The most simple of statistics can be immensely informative.

There are other concerns. The hegemonic position in human geography is now occupied by something that is called "critical human geography." The justifiably lauded *Dictionary of Human Geography* is an exemplar. Nevertheless, the idea needs to be looked at critically. Critical human geography is a highly eclectic mix, drawing variably on some soft critique of capitalism and bits and pieces of "post" thinking that may be no more profound than the use of ideas of identity, social construction, and discourse. It is energized by social oppressions of an economic and cultural sort: the maldistribution of resources and of recognition, therefore. The oppressions of gender, race, ethnicity, development, and poverty attract special attention. The logic and language of class is barely visible. A pluralization of processes rules. The production that the Marxist geographers prioritize has been shunted offstage except as the butt of knee-jerk reactions about "essentialism." Production is not a dirty word, apparently, but it seems that materialism and totalization are. *Tant pis.* I argued in Chapter 9, that progress in human geography is certainly possible. But that does not mean that it is inevitable.

Finally, there are still difficult questions about the relation of human geography to the other social sciences. The question has changed. It is no longer one of finding something that might be of interest; rather, it is now a matter of enhanced awareness of what human geography can offer. The potential is clearly there. Human geographers now have a conceptual and methodological armory that should be of interest to other social scientists. Some inroads have clearly been made. The work of human geographers can now be found in the sociology, economics, and political science journals. But obviously it has not been enough—"obviously" because the other social sciences continue in their half-baked notions of how space should enter into their arguments. There can be few human geographers who have not encountered the egregious ignorance and disciplinary hubris that is out there: throwbacks to the environmental determinism that human geographers rejected almost a century ago; disciplinary imperialism that has not taken the trouble to find out what human geographers have already written on a topic as in the so-called "new" economic geography of the economists; and continuing attempts to force geography into regression equations in order to evaluate its "effects."

So "making space for human geography" remains a work in progress. Human geography now has the concepts and methods to contribute to an understanding of social life in all its different moments. It has successfully colonized a particular slant on the social whole and so can face the future far more optimistically and confidently than would have been possible in the 1950s. There has never been a shortage of enthusiasm in the field and it has finally borne fruit. How to insinuate ourselves into the day-to-day thinking of the other social sciences remains a major challenge.

NOTE

1. Climatology is equally so, but there is a sharp disparity in the attention that anglophone geographers have devoted to geomorphology and climatology, and the same applies to biogeography.

References

Abler, R. F., Adams, J. S., and Gould, P. R. (1971). *Spatial Organization*. Englewood Cliffs, NJ: Prentice Hall.

Agnew, J. A. (1994). "The Territorial Trap: The Geographical Assumptions of International Relations Theory." *Review of International Political Economy, 1*(1), 53–80.

Agnew, J. A. (2005). "Sovereignty Regimes: Territoriality and State Authority in Contemporary World Politics." *Annals of the Association of American Geographers, 95*(2), 437–461.

Agnew, J. A., and Corbridge, S. (1994). *Mastering Space*. London: Routledge.

Agnew, J. A., Livingstone, D. N., and Rogers, A. (1996). *Human Geography: An Essential Anthology*. Oxford, UK: Blackwell.

Albrow, M. (1974). "Dialectical and Categorical Paradigms of a Science of Society." *Sociological Review, 22*(2), 183–201.

Allen, J. (1983). "Property Relations and Landlordism: A Realist Approach." *Environment and Planning D: Society and Space, 1*(2), 191–203.

Allen, J. (2003). *Lost Geographies of Power*. Oxford, UK: Blackwell.

Allen, J. (2010). "Powerful City Networks: More Than Connections, Less Than Domination and Control." *Urban Studies, 47*(13), 2895–2911.

Allen, J., and Cochrane, A. (2007). "Beyond the Territorial Fix: Regional Assemblages, Politics and Power." *Regional Studies, 41*(9), 1161–1175.

Allen, J., and Cochrane, A. (2012). "Assemblages of State Power: Topological Shifts in the Organization of Government and Politics." *Antipode, 42*(5), 1071–1089.

Allen, J., Massey, D., and Cochrane, A. (1998). *Rethinking the Region*. London: Routledge.

Almond, G. A., and Verba, S. (1963). *The Civic Culture: Political Attitudes and Democracy in Five Nations*. Princeton, NJ: Princeton University Press.

259

Alonso, W. (1960). "A Theory of the Urban Land Market." *Papers and Proceedings of the Regional Science Association, 6*, 149–158.

Amin, A. (2002). "Spatialities of Globalization." *Environment and Planning A, 34*, 385–399.

Anselin, L. (1995). "Local Indicators of Spatial Association—LISA." *Geographical Analysis, 27*, 93–115.

Banks, A., and Textor, R. B. (1963). *A Cross-Polity Survey.* Cambridge, MA: MIT Press.

Barnes, T. J. (1998). "A History of Regression: Actors, Networks, Machines and Numbers." *Environment and Planning D, 30*, 203–224.

Barnes, T. J. (2000). "Inventing Anglo-American Economic Geography 1889–1960." Chapter 2 in E. Sheppard and T. J. Barnes (Eds.), *A Companion to Economic Geography.* Oxford, UK: Blackwell.

Barnes, T. J. (2001). "Retheorizing Economic Geography: From the Quantitative Revolution to the 'Cultural Turn.'" *Annals of the Association of American Geographers, 91*(3), 546–565.

Barnes, T. J. (2009). "Regional Science." Pp. 638–639 in D. Gregory et al. (Eds.), *The Dictionary of Human Geography.* Oxford, UK: Blackwell.

Barnett, C. B., and Low, M. M. (1996). "Speculating on Theory: Towards a Political Economy of Academic Publishing." *Area, 28*, 13–24.

Bassett, K. (1999). "Is There Progress in Human Geography?: The Problem of Progress in the Light of Recent Work in the Philosophy and Sociology of Science." *Progress in Human Geography, 23*(1), 27–47.

Bassett, T. J. (1988). "The Political Ecology of Peasant–Herder Conflicts in the Northern Ivory Coast." *Annals of the Association of American Geographers, 78*(3), 453–472.

Bassett, T. J., and Zueli, K. B. (2000). "Environmental Discourses and the Ivorian Savanna." *Annals of the Association of American Geographers, 90*(1), 67–95.

Bell, D. (1960). *The End of Ideology.* Glencoe, IL: Free Press.

Berger, P. L., and Luckmann, T. (1967). *The Social Construction of Reality.* Garden City, NY: Doubleday.

Berry, B. J. L. (1964). "Cities as Systems within Systems of Cities." *Papers and Proceedings of the Regional Science Association, 13*, 147–163.

Berry, B. J. L. (1967). *The Geography of Market Centers and Retail Distribution.* Englewood Cliffs, NJ: Prentice Hall.

Berry, B. J. L., Barnum, H. G., and Tennant, R. J. (1962). "Retail Location and Consumer Behavior." *Papers and Proceedings of the Regional Science Association, 9*, 65–106.

Bertaux, D. (1981). *Biography and Society.* London and Beverly Hills, CA: Sage.

Bhaskar, R. (1979). *The Possibility of Naturalism.* Atlantic Highlands, NJ: Humanities Press.

Bowman, I. (1928; rev. ed.). *The New World.* New York: World Book.

Bowman, I. (1942). "Geography Versus Geopolitics." *Geographical Review, 32*(4), 646–658.

Boyle, M. (2001). "Towards a (Re)theorization of the Historical Geography of Nationalism in Diasporas: The Irish Diaspora as an Exemplar." *International Journal of Population Geography, 7*, 429–446.

Brenner, N. (2004). *New State Spaces.* Oxford: Oxford University Press.

Brenner, R. (1977). "The Origins of Capitalist Development: A Critique of Neo-Smithian Marxism." *New Left Review*, No. 104, 25–92.

Brown, L. A. (1968). *Diffusion Dynamics: A Review and Revision of the Quantitative Theory of the Spatial Diffusion of Innovation* [Lund Studies in Geography, Series B, Human Geography 29]. Lund, Sweden: Gleerup.

Brown, L. A. (1975). "The Market and Infrastructure Context of Adoption: A Spatial Perspective on the Diffusion of Innovation." *Economic Geography, 51,* 185–216.

Brown, L. A. (1981). *Innovation Diffusion: A New Perspective.* London: Methuen.

Bunge, W. (1962). *Theoretical Geography* [Lund Studies in Geography, Series C, General and Mathematical Geography 1]. Lund, Sweden: Gleerup.

Bunge, W. (1964). *Theoretical Geography* (revised and enlarged edition). [Lund Studies in Geography, Series C, General and Mathematical Geography 1]. Lund, Sweden: Gleerup.

Bunge, W. (1971). *Fitzgerald: Geography of a Revolution.* Cambridge, MA: Schenkman.

Callinicos, A. (1988). "Exception or Symptom?: The British Crisis and the World System." *New Left Review*, No. 169, 97–106.

Callinicos, A. (1989). *Against Postmodernism.* London: St. Martin's Press.

Callinicos, A. (2004). "Marxism and the International." *British Journal of Politics and International Relations, 6*(3), 426–433.

Capel, H. (1981). "Institutionalization of Geography and Strategies of Change." Pp. 37–69 in D. Stoddart (Ed.), *Geography, Ideology and Social Concern.* Oxford, UK: Blackwell.

Carney, J. (2002). *Black Rice.* Cambridge MA: Harvard University Press.

Carson, R. (1962). *Silent Spring.* Boston: Houghton Mifflin.

Carswell, G. (2006). "Multiple Historical Geographies: Responses and Resistance to Colonial Conservation Schemes in East Africa." *Journal of Historical Geography, 32,* 398–421.

Casetti, E. (1972). "Generating Models by the Expansion Method: Applications to Geographical Research." *Geographical Analysis, 4*(1), 82–91.

Casetti, E., and Semple, R. K. (1968). *A Method for Stepwise Separation of Spatial Trends* (Discussion Paper No. 11). Ann Arbor: Michigan Inter-University Community of Mathematical Geographers.

Casetti, E., and Semple, R. K. (1969). "Concerning the Testing of Spatial Diffusion Hypotheses." *Geographical Analysis, 1,* 254–259.

Cash, W. J. (1954). *The Mind of the South.* New York: Doubleday Anchor.

Castells, M. (1983). "Crisis, Planning and the Quality of Life: Managing the New Historical Relationships Between Space and Society." *Environment and Planning D, 1*(1), 3–21.

Castree, N., and Braun, B. (2001). *Social Nature.* Oxford, UK: Blackwell.

Chisholm, G. C. (1889). *Handbook of Commercial Geography.* London and New York: Longman, Green.

Chisholm, G. C. (1908). "The Meaning and Scope of Geography." *Scottish Geographical Magazine, 24,* 561–575.

Chisholm, M. (1962). *Rural Settlement and Land Use.* London: Hutchinson.

Chorley, R. J. (1995). "Haggett's Cambridge: 1957–1966." Chapter 17 in A. D. Cliff et al. (Eds.), *Diffusing Geography: Essays for Peter Haggett.* Oxford, UK: Blackwell.

Chorley, R. J., and Haggett, P. (Eds.) (1967). *Models in Geography*. London: Methuen.

Christaller, W. (1933). *Die zentralen Orte in Süddeutschland*. Jena, Germany: Gustav Fischer. (Translated (in part), by Carlisle W. Baskin, as *Central Places in Southern Germany*. Englewood Cliffs, NJ: Prentice Hall, 1966.)

Christopherson, S. (1993). "Market Rules and Territorial Outcomes: The Case of the United States." *International Journal of Urban and Regional Research, 17*(2), 274–288.

Christopherson, S. (2002). "Why Do National Labor Market Practices Continue to Diverge in a Global Economy?." *Economic Geography, 78*, 1–20.

Christopherson, S. (2007). "Barriers to 'US Style' Lean Retailing: The Case of Wal-Mart's Failure in Germany." *Journal of Economic Geography, 7*, 451–469.

Clark, A. H. (1949). *The Invasion of New Zealand by People, Plants and Animals: The South Island*. New Brunswick: Rutgers University Press.

Clark, G. L. (1981). "The Employment Relation and Spatial Division of Labor: A Hypothesis." *Annals of the Association of American Geographers, 71*(3), 412–424.

Clark, G. L. (1993). "Global Interdependence and Regional Development: Business Linkages and Corporate Governance in a World of Financial Risk." *Transactions of the Institute of British Geographers, NS 18*(3), 309–326.

Clark, G. L. (1994). "Strategy and Structure: Corporate Restructuring and the Scope and Characteristics of Sunk Costs." *Environment and Planning A, 26,* 9–32.

Clark, G. L., and Wrigley, N. (1995). "Sunk Costs: A Framework for Economic Geography." *Transactions of the Institute of British Geographers, NS 20*(2), 204–223.

Clark, G. L., and, O'Connor, K. (1997). "The Informational Content of Financial Products and the Spatial Structure of Global Finance." Pp. 89–114 in K. R. Cox (Ed.), *Spaces of Globalization*. New York: Guilford Press.

Clastres, P. (1977). *Society Against the State*. New York: Urizen Books.

Cliff, A. D., Haggett, P., and Ord, J. K. (1986). *Spatial Aspects of Influenza Epidemics*. London: Pion.

Cloke, P., and Johnston, R. J. (Eds.) (2005). *Spaces of Geographical Thought*. London: Sage.

Coates, B., Johnston, R. J., and Knox, P. L. (1977). *Geography and Inequality*. Oxford, UK: Oxford University Press.

Colby, C. (1933). "Centrifugal and Centripetal Forces in Urban Geography." *Annals of the Association of American Geographers, 23*(1), 1–20.

Cole, J. P., and King, C. A. M. (1968). *Quantitative Geography: Techniques and Theories in Geography*. New York: Wiley.

Coleman, M. (2007). "A Geopolitics of Engagement: Neoliberalism, the War on Terrorism, and the Reconfiguration of US Immigration Enforcement." *Geopolitics, 12*(4), 607–634.

Colenutt, B. (1970). "Poverty and Inequality in American Cities." *Antipode, 2*(2), 55–60.

Cooke, P. (Ed.) (1989). *Localities*. London: Unwin Hyman.

Corbridge, S. (1986). *Capitalist World Development*. Totowa, NJ: Rowman and Littlefield.

Cornish, V. (1923). *The Great Capitals: An Historical Geography*. London: Methuen.

Corry, M., and Sugiura, Y. (1999). *Exceptionalism in Geography Revisited: Papers* (Discussion Paper No. 45). Iowa City: University of Iowa, Department of Geography.

Cosgrove, D. (2006). "Modernity, Community and the Landscape Idea." *Journal of Material Culture, 11*(1)-2, 49–66.

Cox, K. R. (1965). "The Application of Linear Programming to Geographic Problems." *Tijdschrift voor Economische en Sociale Geografie, 56*(6), 228–236.

Cox, K. R. (1969a). "Spatial Structuring of Information Flow and Partisan Attitudes." Pp. 157–185 in M. Dogan and S. Rokkan (Eds.), *Quantitative Ecological Analysis in the Social Sciences.* Cambridge, MA: MIT Press.

Cox, K. R. (1969b). "The Voting Decision in a Spatial Context." Pp. 81–117 in C. Board, R. J. Chorley, P. Haggett, and D. R. Stoddart (Eds.), *Progress in Geography I.* London: Edward Arnold.

Cox, K. R. (1970). "Residential Relocation and Political Behavior: Conceptual Model and Empirical Tests." *Acta Sociologica, 13*(6), 40–53.

Cox, K. R. (1972a). *Man, Location and Behavior.* New York: Wiley.

Cox, K. R. (1972b). "The Spatial Components of Urban Voting Response Surfaces." *Economic Geography, 47*(1), 27–35.

Cox, K. R. (1981). "Bourgeois Thought and the Behavioral Geography Debate." Pp. 256–279 in K. R. Cox and R. G. Golledge (Eds.), *Behavioral Geography Revisited.* London: Methuen.

Cox, K. R. (1998). "Spaces of Dependence, Spaces of Engagement and the Politics of Scale, or: Looking for Local Politics." *Political Geography, 17*(1), 1–24.

Cox, K. R. (2006). "Bringing Physical Geography into Geographic Thought Courses." *Journal of Geography in Higher Education, 30*(3), 373–388.

Cox, K. R. (2013a). "Notes on a Brief Encounter: Critical Realism, Historical Materialism and Human Geography." *Dialogues in Human Geography, 3*(1), 3–21.

Cox, K. R. (2013b). "Territory, Scale and Why Capitalism Matters." *Territory, Politics and Governance, 1*(1).

Cox, K. R., and Johnston, R. J. (Eds.) (1982). *Conflict, Politics and the Urban Scene.* Harlow, Essex: Longman.

Cox, K. R., and Mair, A. (1988). "Locality and Community in the Politics of Local Economic Development." *Annals of the Association of American Geographers, 78*(2), 307–325.

Cox, K. R., and Wood, A. M. (1997). "Competition and Cooperation in Mediating the Global: The Case of Local Economic Development." *Competition and Change, 2*(1), 65–94.

Cox, K. R., and Zannaras, G. (1973). "Designative Perceptions of Macro Spaces: Concepts, A Methodology and Application." Pp. 162–178 in R. Downs and D. Stea (Eds.), *Cognitive Mapping: Images of Spatial Environments.* New York: Aldine.

Curry, L. (1964). "The Random Space Economy: An Exploration in Settlement Theory." *Annals of the Association of American Geographers, 38*, 138–146.

Curry, L. (1966). "Chance and Landscape." Pp. 40–55 in J. W. House (Ed.), *Northern Geographical Essays.* Newcastle Upon Tyne, UK: Oriel.

Curry, L. (1972). "A Spatial Analysis of Gravity Flows." *Regional Studies, 6*, 131–147.

Cutright, P. (1963). "National Political Development: Measurement and Analysis." *American Sociological Review, 28*, 253–264.

Cutright, P. (1965). "Political Structure, Economic Development and National Security Programs." *American Journal of Sociology, 70*(5), 537–550.

Cutright, P., and Wiley, J. A. (1969). "Modernization and Political Representation: 1927–1966." *Studies in Comparative International Development, 5*, 23–44.

Cybriwsky, R. A. (1978). "Social Aspects of Neighborhood Change." *Annals of the Association of American Geographers, 68*(1), 17–33.

Dacey, M. F. (1968). "An Empirical Study of the Areal Distribution of Houses in Puerto Rico." *Transactions of the Institute of British Geographer, 45*, 51–70.

Darby, H. C. (1951). "The Changing English Landscape." *Geographical Journal, 117*, 377–398.

Darby, H. C. (1953). "On the Relations of Geography and History." *Transactions of the Institute of British Geographers, 19*, 1–11.

Darby, H. C. (1956). *The Draining of the Fens*. Cambridge, UK: Cambridge University Press.

Darby, H. C. (Ed.) (1973). *A New Historical Geography of England*. Cambridge, UK: Cambridge University Press.

David, T. (1958). "Against Geography." *Universities Quarterly, 12*(3), 261–273.

Davis, D. K. (2007). *Resurrecting the Granary of Rome*. Athens: Ohio University Press.

Davis, W. M. (1906). "An Inductive Study of the Content of Geography." *Bulletin of the American Geographical Society, 38*, 67–84.

Dicken, P. (2004). "Geographers and Globalization: (Yet) Another Boat Missed?." *Transactions of the Institute of British Geographers, NS 29*(1), 5–26.

Dickinson, R. E. (1934). "Markets and Market Areas of East Anglia." *Economic Geography, 10*(2), 172–182.

Dickinson, R. E., and Howarth, O. J. R. (1933). *The Making of Geography*. Oxford: Clarendon Press.

Dogan, M., and Rokkan, S. (Eds.) (1969). *Quantitative Ecological Analysis in the Social Sciences*. Cambridge, MA: MIT Press.

Dorling, D. (1995). *A New Social Atlas of Britain*. London: John Wiley.

Dorling, D., and Thomas, B. (2011). *Bankrupt Britain: An Atlas of Social Change*. Bristol: Policy Press.

Downs, R. B. (1956). *Books That Changed the World*. New York: New American Library.

Driver, F. (1992). "Geography's Empire: Histories of Geographical Knowledge." *Environment and Planning D, 10*(1), 23–40.

Driver, F. (2003). "On Geography as a Visual Discipline." *Antipode, 35*(2), 227–231.

Duhl, L. J., and Steetle, N. J. (1969). "Newark: Community or Chaos—A Case Study of the Medical School Controversy." *Journal of Applied Behavioral Science, 5*(4), 537–572.

Duncan, J. S., and Ley, D. (1982). "Structural Marxism and Human Geography: A Critical Assessment." *Annals of the Association of American Geographers, 72*, 30–59.

Duncan, O. D., Cuzzort, R. P., and Duncan, B. (1961). *Statistical Geography*. Glencoe, IL: Free Press.

Eagleton, T. (1995). "Where do Postmodernists Come From?." *Monthly Review, 47*(3), 59–71.

East, W. G., and Moodie, A. E. (1956). *The Changing World: Studies in Political Geography*. London: George Harrap.

Easterly, W., and Levine, R. (2003). "Tropics, Germs, and Crops: How Endowments Influence Economic Development." *Journal of Monetary Economics, 50*, 3–39.

Ehrlich, P. R., and Ehrlich, A. H. (1968). *The Population Bomb*. New York: Sierra Club and Ballantine Books.

England, K. V. L. (1993). "Suburban Pink Collar Ghettoes: The Spatial Entrapment of Women?." *Annals of the Association of American Geographers, 83*(2), 225–242.

England, K. V. L. (1994). "Getting Personal: Reflexivity, Positionality and Feminist Research." *Professional Geographer, 46*(1), 80–89.

Entrikin, N. (1976). "Contemporary Humanism in Geography." *Annals of the Association of American Geographers, 66*, 615–632.

Fairgrieve, J. (1915). *Geography and World Power*. London: University of London Press.

Fall, J. (2005). "Michel Foucault and Francophone Geography." *EspaceTemps* (www.espacestemps.net/document1540.html; last accessed June 1, 2012).

Fall, J. (2010). "Artificial States? On the Enduring Myth of Natural Borders." *Political Geography, 29*(3), 140–147.

Farmer, B. H. (1957). *Pioneer Colonization in Ceylon*. London: Oxford University Press.

Fawcett, C. B. (1917). "The Natural Divisions of England." *Geographical Journal, 49*(2), 124–135.

Fitzgerald, W. (1945). *The New Europe*. London: Methuen.

Foord, J., and Gregson, N. (1986). "Patriarchy: Towards a Reconceptualization." *Antipode, 18*(2), 186–211.

Forde, C. D. (1925). "Values in Human Geography." *Geography, 13*(3), 216–221.

Forde, C. D. (1934). *Habitat, Economy and Society: A Geographical Introduction to Ethnology*. London: Methuen.

Fotheringham, A. S. (1981a). "Spatial Structure and Distance-Decay Parameters." *Annals of the Association of American Geographers, 71*(3), 425–436.

Fotheringham, A. S. (1981b). "Trends in Quantitative Methods 1: Stressing the Local." *Progress in Human Geography, 21*(1), 88–96.

Frank, A. G. (1972). "Sociology of Development and Underdevelopment of Sociology." Chapter 11 in J. D. Cockcroft, A. G. Frank, and D. L. Johnson (Eds.), *Dependence and Underdevelopment*. Garden City, NY: Doubleday Anchor.

Friedman, T. L. (2007). *The World is Flat*. New York: Picador.

Garland, J. H. (1955). *The North American Midwest: A Regional Geography*. New York: Wiley.

Garrison, W. L. (1959–1960). "The Spatial Structure of the Economy." *Annals of the Association of American Geographers, 49*, 232–239, 471–482; *50*, 357–373.

Garrison, W. L. (1960). "Connectivity of the Interstate Highway System." *Papers and Proceedings of the Regional Science Association, 6*, 127–137.

Garrison, W. L., Berry, B. J. L., Marble, D. F., Nyestuen, J. N., and Morrill, R. L. (1959). *Studies of Highway Development and Geographic Change*. Seattle: University of Washington Press.

Garrison, W. L., and Marble, D. F. (1965). *A Prolegomenon to the Forecasting of*

Transportation Development. Evanston, IL: Northwestern University, Transportation Center.

Gendron, R., and Domhoff, G. W. (2008). *The Leftmost City*. Boulder, CO: Westview Press.

Gertler, M. (1997). "Between the Global and the Local: The Spatial Limits to Productive Capital." Chapter 2 in K. R. Cox (Ed.), *Spaces of Globalization*. New York: Guilford Press.

Giddens, A. (1981). *Contemporary Critique of Historical Materialism*. Berkeley and Los Angeles: University of California Press.

Gidwani, V., and Sivaramakrishnan, K. (2003). "Circular Migration and Spaces of Cultural Assertion." *Annals of the Association of American Geographers, 93*(1), 186–213.

Golledge, R. G. (1969). "The Geographical Relevance of some Learning Theories." Pp. 101–145 in K. R. Cox and R. G. Golledge (Eds.), *Behavioral Geography: A Symposium*. Evanston, IL: Northwestern Studies in Geography, 17.

Golledge, R. G. (1970). "Some Equilibrium Models of Consumer Behavior." *Economic Geography, 46*, 417–424.

Gough, J. (1991). "Structure, System and Contradiction in the Capitalist Space Economy." *Environment and Planning D: Society and Space, 9*, 433–449.

Gough, J. (2002). "Neo-Liberalism and Socialization in the Contemporary City: Opposites, Complements and Instabilities." *Antipode, 34*(3), 405–426.

Gough, J., and Eisenschnitz, A. (1997). "The Division of Labor, Capitalism and Socialism: An Alternative to Sayer." *International Journal of Urban and Regional Research, 21*(1), 23–37.

Gould, P. R. (1970). "Is Statistix Inferens the Geographical Name for a Wild Goose?" *Economic Geography (Supplement)* 46, 439–448.

Gould, P. R. (1975). "Acquiring Spatial Information." *Economic Geography, 51*(2), 87–99.

Gould, P. R. (1979). "Geography 1957–77: The Augean Period." *Annals of the Association of American Geographers, 69*(1), 139–151.

Gould, P. R., and White, R. (1974). *Mental Maps*. Harmondsworth, Middlesex, UK: Penguin.

Gourou, P. (1953). *The Tropical World*. London: Longman.

Graham, B. J. (1994). "The Search for Common Ground: Estyn Evans's Ireland." *Transactions of the Institute of British Geographers, NS 19*(2), 183–201.

Gravier, J.-F. (1947). *Paris et le désert français*. Paris: Portulan.

Green, F. H. W. (1950). "Urban Hinterlands in England and Wales: An Analysis of Bus Services." *Geographical Journal, 96*, 64–81.

Gregory, D. (1978). *Ideology, Science and Human Geography*. London: Hutchinson.

Gregory, D. (1980). "The Ideology of Control: Systems Theory and Geography." *Tijdschrift voor Economische en Sociale Geografie, 71*, 327–342.

Gregory, D. (1981). "Human Agency and Human Geography." *Transactions of the Institute of British Geographers, NS 6*, 1–18.

Gregory, D. (1985). "Suspended Animation: The Stasis of Diffusion Theory." Chapter 13 in D. Gregory and J. Urry (Eds.), *Social Relations and Spatial Structures*. London: Macmillan.

Gregory, D. (1990). "*Chinatown*, Part 3? Soja and the Missing Spaces of Social Theory." *Strategies, 3*, 40–104.

Gregory, D. (1994). *Geographical Imaginations*. Oxford, UK: Blackwell.

Gregory, D. (1998). "Power, Knowledge and Geography." *Geographische Zeitschrift, 86*, 70–93.

Gregory, D. (2004). *The Colonial Present*. Oxford, UK: Blackwell.

Gregory, D. (2009). "Space." Pp. 707–710 in D. Gregory et al. (Eds.), *The Dictionary of Human Geography*. Oxford, UK: Blackwell.

Gregory, D., and Urry, J. (1985). *Social Relations and Spatial Structure*. London: Macmillan.

Gregson, N. (1987). "Structuration Theory: Some Thoughts on the Possibilities for Empirical Research." *Environment and Planning D, 5*(1), 73–91.

Gregson, N. (2005). "Agency-Structure." Chapter 2 in P. Cloke and R. J. Johnston (Eds.), *Spaces of Geographic Thought*. London: Sage.

Grigg, D. (1979). "Ester Boserup's Theory of Agrarian Change." *Progress in Human Geography, 3*(1), 64–84.

Gruffudd, P. (1994). "Back to the Land: Historiography, Rurality and the Nation in Interwar Wales." *Transactions of the Institute of British Geographers, NS 19*(1), 61–77.

Gunn, R. (1989). "Marxism and Philosophy." *Capital and Class*, No. 37, 87–116.

Habermas, J. (1970). "Technology and Science as 'Ideology.'" Chapter 6 in *Toward a Rational Society: Student Protest, Science, and Politics*. Boston: Beacon Press.

Hägerstrand, T. (1957). "Migration and Area: Survey and Sample of Swedish Migration Fields and Hypothetical Considerations on their Genesis." Pp. 27–158 in D. Hannerberg, T. Hägerstrand, and B. Odeving (Eds.), *Migration in Sweden: A symposium* [Lund Studies in Geography, Series B, Human Geography 13]. Lund, Sweden: Gleerup.

Hägerstrand, T. (1965). "A Monte Carlo Approach to Diffusion." *European Journal of Sociology, 6*, 43–67.

Hägerstrand, T. (1967). *Innovation Diffusion as a Spatial Process*. Chicago: University of Chicago Press.

Hägerstrand, T. (1970). "What about People in Regional Science?" *Papers and Proceedings of the Regional Science Association, 24*, 7–21.

Hägerstrand, T. (1973). "The Domain of Human Geography." Chapter 4 in R. J. Chorley (Ed.), *Directions in Geography*. London: Methuen.

Haggett, P. (1961). "Land Use and Sediment Yield in an Old Plantation Tract of the Serra do Mar, Brazil." *Geographical Journal, 127*, 50–62.

Haggett, P. (1964). "Regional and Local Components in the Distribution of Forested Areas in Southeast Brazil: A Multivariate Approach." *Geographical Journal, 130*, 365–380.

Haggett, P. (1965). *Locational Analysis in Human Geography*. London: Edward Arnold.

Haggett, P., and Chorley, R. J. (1969). *Network Analysis in Geography*. London: Edward Arnold.

Hanson, S., and Pratt, G. (1988). "Spatial Dimensions of the Gender Division of Labor in a Local Labor Market." *Urban Geography, 9*, 367–378.

Hanson, S., and Pratt, G. (1990). "Geographic Perspectives on the Occupational Segregation of Women." *National Geographic Research, 6*(4), 376–399.

Hanson, S., and Pratt, G. (1991). "Job Search and the Occupational Segregation of Women." *Annals of the Association of American Geographers, 81*(2), 229–253.

Harloe, M., Pickvance, C., and Urry, J. (1990). *Place, Policy and Politics*. London: Unwin Hyman.

Harris, C. D. (1954). "The Market as a Factor in the Localization of Industry in the United States." *Annals of the Association of American Geographers, 44,* 315–348.

Harris, R. C. (1978). "The Historical Mind and the Practice of Geography." Chapter 8 in D. Ley and M. Samuels (Eds.), *Humanistic Geography*. Chicago: Maaroufa.

Hartshorne, R. (1927). "Location as a Factor in Geography." *Annals of the Association of American Geographers, 17*(2), 92–99.

Hartshorne, R. (1939). *The Nature of Geography*. Washington, DC: Association of American Geographers.

Hartshorne, R. (1959). *Perspective on the Nature of Geography*. Chicago: Rand McNally.

Harvey, D. (1966). "Geographical Processes and the Analysis of Point Patterns: Testing Models of Diffusion by Quadrat Sampling." *Transactions of the Institute of British Geographers, 40,* 81–95.

Harvey, D. (1971). "Social Processes, Spatial Form and the Redistribution of Real Income in an Urban System." Pp. 267–300 in M. Chisholm (Ed.), *Regional Forecasting*. London: Butterworth.

Harvey, D. (1973). *Social Justice and the City*. Baltimore, MD: Johns Hopkins University Press.

Harvey, D. (1974). "Population, Resources and the Ideology of Science." *Economic Geography, 50,* 256–267.

Harvey, D. (1975). "The Geography of Capitalist Accumulation: A Reconstruction of the Marxian Theory." *Antipode, 7*(2), 9–21.

Harvey, D. (1978). "Labor, Capital and Class Struggle around the Built Environment in Advanced Capitalist Societies." Chapter 1 in K. R. Cox (Ed.), *Urbanization and Conflict in Market Societies*. Chicago: Maaroufa.

Harvey, D. (1982). *The Limits to Capital*. Oxford, UK: Blackwell.

Harvey, D. (1985a). "The Geopolitics of Capitalism." Chapter 7 in D. Gregory and J. Urry (Eds.), *Social Relations and Spatial Structures*. London: Macmillan.

Harvey, D. (1985b). "The Place of Urban Politics in the Geography of Uneven Capitalist Development." Chapter 6 in *The Urbanization of Capital*. Baltimore, MD: Johns Hopkins University Press.

Harvey, D. (1987). "Three Myths in Search of a Reality in Urban Studies." *Environment and Planning A, 5*(4), 367–376.

Harvey, D. (1989). *The Condition of Postmodernity*. Oxford, UK: Blackwell.

Harvey, D. (1992). "Postmodern Morality Plays." *Antipode, 24*(4), 300–326.

Harvey, D. (1996). *Justice, Nature and the Geography of Difference*. Oxford, UK: Blackwell.

Harvey, D. (2003). *The New Imperialism*. Oxford, UK: Oxford University Press.

Harvey, D. (2005a). "The Sociological and Geographical Imaginations." *International Journal of Politics, Culture and Society, 18,* 211–255.

Harvey, D. (2005b). *A Brief History of Neoliberalism*. Oxford, UK: Oxford University Press.

Hepple, L. (2002). "Socialist Geography in England: J F Horrabin and a Workers' Economic and Political Geography." *Antipode, 31*(1), 80–109.

Herbert, S. (1997). *Policing Space: Territoriality and the Los Angeles Police Department.* Minneapolis: University of Minnesota Press, 1997.

Herbert, S. (2000). "For Ethnography." *Progress in Human Geography, 24*(4), 550–568.

Hirst, P., and, G. Thompson (1996). *Globalization in Question.* Cambridge, UK: Polity Press.

Hobsbawm, E. (1996). "Identity Politics and the Left." *New Left Review, 217,* 38–47.

Hodder, I., and Orton, C. (1976). *Spatial Analysis in Archaeology.* Cambridge, UK: Cambridge University Press.

Horrabin, J. F. (1923). *An Outline of Economic Geography.* London: The Plebs League.

Horrabin, J. F. (1934). *Atlas of Current Affairs.* London: Victor Gollancz.

Hoskins, W. G. (1955). *The Making of the English Landscape.* Leicester: Leicester University Press.

Houston, J. M. (1959). "Land Use and Society in the Plain of Valencia." In R. Miller and J. W. Watson (Eds.), *Geographical Essays in Honor of Alan G. Ogilvie.* London: Thomas Nelson.

Huntington, E. (1915). *Civilization and Climate.* New Haven: Yale University Press.

Huntington, E. (1945). *Mainsprings of Civilization.* New York: Wiley.

Jansson, D. R. (2003). "Internal Orientalism in America: W.J. Cash's *The Mind of the South* and the Spatial Construction of American National Identity." *Political Geography, 22,* 293–316.

Jarosz, L. (1992). "Constructing the Dark Continent: Metaphor as Geographic Representation of Africa." *Geografiska Annaler, Series B, Human Geography, 74*(2), 105–115.

Jewitt, S. (1995). "Europe's 'Others'? Forestry Policy and Practices in Colonial and Postcolonial India." *Environment and Planning D, 13*(1), 67–90.

Johnston, R. J. (2001). "Robert E Dickinson and the Growth of Urban Geography: An Evaluation." *Urban Geography, 22*(8), 702–736.

Johnston, R. J. (2002). "Manipulating Maps and Winning Elections: Measuring the Impact of Malapportionment and Gerrymandering." *Political Geography, 21*(1), 1–32.

Johnston, R. J. (2005). "Geography—Coming Apart at the Seams?" Chapter 1 in N. Castree, A. Rogers, and D. Sherman (Eds.), *Questioning Geography.* Oxford, UK: Blackwell.

Johnston, R. J., Gregory, D., Pratt, G., and Watts, M. (Eds.) (2000). *The Dictionary of Human Geography.* Oxford, UK: Blackwell.

Johnston, R. J., and Sidaway, J. D. (2004). *Geography and Geographers.* London: Arnold.

Jones, K. (1991). "Specifying and Estimating Multi-Level Models for Geographical Research." *Transactions of the Institute of British Geographers, NS 16*(2), 148–159.

Jones, K., Johnston, R. J., and Pattie, C. J. (1992). "People, Places and Regions: Exploring the Use of Multi-Level Modeling in the Analysis of Electoral Data." *British Journal of Political Science, 22*(3), 343–380.

Kaplan, R. (2012). *The Revenge of Geography: What the Map Tells Us About Coming Conflicts and the Battle Against Fate.* New York: Random House.

Katz, C. (2001). "Vagabond Capitalism and the Necessity of Social Reproduction." *Antipode, 33*(4), 709–728.

Kearns, G. (1984). "Closed Space and Political Practice: Frederick Jackson Turner and Halford Mackinder." *Environment and Planning D, 2*(1), 23–34.

King, L. J. (1969). *Statistical Analysis in Geography*. Englewood Cliffs, NJ: Prentice Hall.

Kirk, W. (1963). "Problems of Geography." *Geography, 48*, 357–371.

Knox, P. (1975). *Social Well-Being: A Spatial Perspective*. London: Oxford University Press.

Knox, P., Agnew, J. A., and McCarthy, L. (2008). *Geography of the World Economy*. New York: Oxford University Press.

Kuhn, T. S. (1962). *The Structure of Scientific Revolutions*. Chicago: University of Chicago Press.

Lambert, J. M., Jennings, J. N., Smith, C. T., and Godwin, H. (1960). *The Making of the Broads*. London: The Royal Geographical Society.

Lee, R., and Wills, J. (Eds.) (1997). *Geographies of Economies*. London: Arnold.

Lévy, J., and Lussault, M. (Eds.) (2003). *Dictionnaire de la géographie et de l'espace des sociétés*. Paris: Belin.

Ley, D. (1977). "Social Geography and the Taken-for-Granted World." *Transactions of the Institute of British Geographers, NS 2*, 498–512.

Ley, D. (1980). "Liberal Ideology and the Post-Industrial City." *Annals of the Association of American Geographers, 70*, 238–258.

Ley, D. (1981). "Behavioral Geography and the Philosophies of Meaning." Chapter 9 in K. R. Cox and R. G. Golledge (Eds.), *Behavioral Problems in Geography Revisited*. London: Methuen.

Ley, D. (1988). "Interpretive Social Research in the Inner City." Chapter 8 in J. Eyles (Ed.), *Research in Human Geography*. Oxford, UK: Blackwell.

Ley, D., and Samuels, M. (1978). "Introduction: Contexts of Modern Humanism in Geography." In D. Ley and M. Samuels (Eds.), *Humanistic Geography*. Chicago: Maaroufa.

Lipietz, A. (1986). "New Tendencies in the International Division of Labor: Regimes of Accumulation and Modes of Regulation," Chapter 2 in A. J. Scott and M. Storper (Eds.), *Production, Work, Territory*. London: Allen & Unwin.

Livingstone, D. N. (1992). *The Geographical Tradition*. Oxford, UK: Blackwell.

Logan, W. S. (1968). "The Changing Landscape Significance of the Victoria-South Australia Boundary." *Annals of the Association of American Geographers, 58*(1), 128–154.

Lösch, A. (1954). *The Economics of Location*. New Haven: Yale University Press.

Lovering, J. (1987). "Militarism, Capitalism and the Nation-State: Toward a Realist Synthesis." *Environment and Planning D, 5*, 283–302.

Lovering, J. (1990). "Fordism's Unknown Successor: A Comment on Scott's Theory of Flexible Accumulation and the Re-Emergence of Regional Economies." *International Journal of Urban and Regional Research, 14*(1), 159–174.

Lövgren, E. (1957). "Mutual Relations between Migration Fields: A Circulation Analysis." Pp. 159–169 in D. Hannerberg, T. Hägerstrand, and B. Odeving (Eds.), *Migration in Sweden: A symposium* [Lund Studies in Geography, Series B, Human Geography 13]. Lund, Sweden: Gleerup.

Lowenthal, D. (1961). "Geography, Experience and Imagination: Towards a Geographical Epistemology." *Annals of the Association of American Geographers, 51*(3), 241–260.

Lowenthal, D. (1975). "Past Time, Present Place: Landscape and Memory." *Geographical Review, 65*(1), 1-36.

Lowenthal, D. (1991). "British National Identity and the English Landscape." *Rural History, 2*, 205-230.

Mackay, R. (1958). "The Interactance Hypothesis and Boundaries in Canada: A Preliminary Study." *Canadian Geographer, 11*, 1-8.

Mackinder, H. J. (1907). *Britain and the British Seas.* Oxford, UK: Clarendon Press.

Mackinder, H. J. (1919). *Democratic Ideals and Reality.* London: Constable.

Mair, A. (1986). "Thomas Kuhn and Understanding Geography." *Progress in Human Geography, 10*, 345-370.

Mamdani, M. (1996). *Citizen and Subject.* Princeton, NJ: Princeton University Press.

Mamdani, M. (2001). "Beyond Settler and Native as Political Identities: Overcoming the Political Legacy of Colonialism." *Comparative Studies in Society and History, 43*(4), 651-664.

Mandel, E. (1986). *The Meaning of the Second World War.* London: Verso.

Manley, G. (1952). *Climate and the British Scene.* London: Collins.

Marchand, B. (2007). "The Fight against the Big City: Urbaphobia Since 200 Years." *Urban Research Network Newsletter*, No. 1, 1-26.

Marchand, B., and Cavin, J. S. (2007). "Anti-Urban Ideologies and Planning in France and Switzerland." *Planning Perspectives, 22*, 29-53.

Marston, S., Jones, J. P., and Woodward, K. (2005). "Human Geography Without Scale." *Transactions of the Institute of British Geographers, NS 30*(4), 416-432.

Martin, R. (1999). "The New 'Geographical Turn' in Economics: Some Critical Reflections." *Cambridge Journal of Economics, 23*, 65-91.

Martin, R. (2010a). "Rethinking Regional Path Dependence: From Lock-In to Evolution." *Economic Geography, 86*(1), 1-27.

Martin, R. (2010b). "The Place of Path Dependence in an Evolutionary Perspective on the Economic Landscape." Pp. 62-92 in R. Boschma and R. Martin (Eds.), *Handbook of Evolutionary Economic Geography.* Cheltenham, UK: Edward Elgar.

Martin, R., and Sunley, P. (1996). "Paul Krugman's Geographical Economics and Its Implications for Regional Development Theory: A Re-Assessment." *Economic Geography, 72*(3), 259-292.

Martin, R., and Sunley, P. (2007). "Complexity Thinking and Evolutionary Economic Geography." *Journal of Economic Geography, 7*(5), 573-601.

Martin, R. L., Sunley, P., and Wills, J. (1996). *Union Retreat and the Regions.* London: Routledge.

Marx, K. (1857-1858; reprinted 1973). *Grundrisse.* Harmondsworth, Middlesex, UK: Penguin.

Marx, K. (1869; reprinted 1973). "The Eighteenth Brumaire of Louis Napoleon." Pp. 143-249 in D. Fernbach (Ed.), *Karl Marx: Surveys from Exile.* Harmondsworth, Middlesex, UK: Penguin Books.

Marx, K., and Engels, F. (1846; reprinted 1978). *The German Ideology.* New York: International Publishers.

Massey, D. (1979). "In What Sense a Regional Problem?" *Regional Studies, 13*(2), 233-243.

Massey, D. (1983). "Industrial Restructuring as Class Restructuring: Production Decentralization and Local Uniqueness." *Regional Studies, 17*(2), 73-89.

Massey, D. (1984). *Spatial Divisions of Labor.* London: Macmillan.

Massey, D. (1985). "New Directions in Space." Chapter 2 in D. Gregory and J. Urry (Eds.), *Social Relations and Spatial Structures.* London: Macmillan.

Massey, D. (1991). "Flexible Sexism." *Environment and Planning D, 9*(1), 31–57.

Massey, D. (1992). "Politics and Space/Time." *New Left Review, 165,* 65–84.

Massey, D. (1995). "Masculinity, Dualisms and High Technology." *Transactions of the Institute of British Geographers, 20*(4), 487–499.

Massey, D. (1999a). "Spaces of Politics." Chapter 14 in D. Massey, J. Allen, and P. Sarre (Eds.), *Human Geography Today.* Cambridge, MA: Polity Press.

Massey, D. (1999b). "Space-Time, 'Science' and the Relationship between Physical Geography and Human Geography." *Transactions of the Institute of British Geographers, 24,* 261–276.

Massey, D. (2001). "Geography on the Agenda." *Progress in Human Geography, 25*(1), 5–17.

Massey, D. (2007). *World City.* Cambridge, UK: Polity Press.

Massey, D., and Meegan, R. A. (1978). "Industrial Restructuring versus the Cities." *Urban Studies, 15*(3), 273–288.

Massey, D., and Thrift, N. (2003). "The Passion of Place." Chapter 8 in R. J. Johnston and M. Williams (Eds.), *A Century of British Geography.* Oxford, UK: Oxford University Press.

Mayer, H. M. (1964). "Politics and Land Use: The Indiana Shoreline of Lake Michigan." *Annals of the Association of American Geographers, 54*(4), 508–523.

McCarty, H. H., Hook, J. C., and Knos, D. S. (1956). "The Measurement of Association in Industrial Geography." *State University of Iowa, Department of Geography, Report, 1.*

McCarty, H. H., and Salisbury, N. E. (1961). "Visual Comparison of Isopleth Maps as a Means of Determining Correlations Between Spatially Distributed Phenomena." *State University of Iowa, Department of Geography, Report, 3.*

McDowell, L. (1983). "Towards an Understanding of the Gender Division of Urban Space." *Environment and Planning D, 1*(1), 59–72.

McDowell, L. (1986). "Beyond Patriarchy: A Class-Based Explanation of Women's Subordination." *Antipode, 18*(3), 311–321.

McDowell, L. (2004). "Masculinity, Identity and Labour Market Change: Some Reflections on the Implications of Thinking Relationally about Difference and the Politics of Inclusion." *Geografiska Annaler, Series B, Human Geography, 86*(1), 45–56.

McDowell, L., and Massey, D. (1984). "A Woman's Place." Chapter 16 in D. Massey and J. Allen (Eds.), *Geography Matters!* Cambridge, UK: Cambridge University Press.

McLennan, D. (1989). *Marxism, Pluralism and Beyond.* Oxford, UK: Blackwell.

McNally, D. (1995). "Language, History and Class Struggle." *Monthly Review, 47*(3), 13–31.

Meadows, D., Meadows, D. L., Randers, J., and Behrens, W. W. (1972). *The Limits to Growth.* New York: Universe Books.

Meinig, D. (1983). "Geography as an Art." *Transactions of the Institute of British Geographers, 8*(3), 314–328.

Merritt, R. L., and Rokkan, S. (Eds.) (1966). *Comparing Nations: The Use of Quantitative Data in Cross-National Research.* New Haven, CT: Yale University Press.

Mitchell, D. (1996). *The Lie of the Land.* Minneapolis: University of Minneapolis Press.

Mitchell, D. (2000). *Cultural Geography.* Oxford, UK: Blackwell.

Mitchell, J. B. (Ed.) (1962). *Great Britain: Geographical Essays.* Cambridge, UK: Cambridge University Press.

Mitchell, K. (1993). "Multiculturalism, or the United Colors of Capitalism." *Antipode, 25*(4), 263–294.

Mitchell, T. (2002). *The Rule of Experts.* Berkeley and Los Angeles: University of California Press.

Moellering, H., and Tobler, W. R. (1972). "Geographical Variances." *Geographical Analysis, 4*(1), 34–50.

Molotch, H. (1976). "City as a Growth Machine: Toward a Political Economy of Place." *American Journal of Sociology, 82*(2), 309–332.

Monk, J. (2004). "Women, Gender and the Histories of American Geography." *Annals of the Association of American Geographers, 94*(1), 1–22.

Morrill, R. L. (1963). "The Development of Spatial Distributions of Towns in Sweden: An Historical-Predictive Approach." *Annals of the Association of American Geographers, 53*(1), 1–14.

Morrill, R. L. (1965a). "Expansion of the Urban Fringe: A Simulation Experiment." *Papers, Regional Science Association, 15,* 173–183.

Morrill, R. L. (1965b). "The Negro Ghetto: Problems and Alternatives." *Geographical Review, 55,* 339–361.

Moser, C., and Scott, W. (1961). *British Towns: A Statistical Study of their Social and Economic Differences.* London: Oliver and Boyd.

Muth, R. (1962). "The Spatial Structure of the Housing Market." *Papers and Proceedings of the Regional Science Association, 7,* 207–220.

Nah, A. H. (2006). "(Re)Mapping Indigenous 'Race'/Place in Postcolonial Peninsular Malaysia." *Geografiska Annaler, Series B, Human Geography, 88*(3), 285–297.

Naumann, F. (1915). *Mitteleuropa.* Berlin: Reimer.

Nelson, K. (1986). "Labor Demand, Labor Supply and the Suburbanization of Low-Wage Office Work." Pp. 149–71 in A. Scott and M. Storper (Eds.), *Production, work and territory.* Winchester, MA: Allen and Unwin.

Newman, D., and Paasi, A. (1998). "Fences and Neighbors in the Postmodern World: Boundary Narratives in Political Geography." *Progress in Human Geography, 22*(2), 186–207.

Norcliffe, G. B. (1977). *Inferential Statistics for Geographers.* New York: Wiley.

"Obituary for Stanley H. Beaver" (1985). *Transactions of the Institute of British Geographers, NS 10*(4) (1985), 504–506.

O'Dell, A. C. (1956). *Railways and Geography.* London: Hutchinson.

Odell, P. R. (1957). "The Hinterlands of Melton Mowbray and Coalville." *Transactions of the Institute of British Geographers,* No. 23, 175–190.

Odland, J., Casetti, E., and King, L. J. (1973). "Testing Hypotheses of Polarized Growth within a Central Place Hierarchy." *Economic Geography, 49*(1), 74–79.

Openshaw, S. (1984). "Ecological Fallacies and the Analysis of Areal Census Data." *Environment and Planning A, 16,* 17–32.

Page, B. (1996). "Across the Great Divide: Agriculture and Industrial Geography." *Economic Geography, 72*(4), 376–397.

Philo, C. (1992). "Foucault's Geography." *Environment and Planning D: Society and Space, 10*(2), 137–161.

Philo, C. (1994). "History, Geography and the 'Still Greater Mystery' of Historical Geography." Chapter 10 in D. Gregory, R. Martin, and G. Smith (Eds.), *Human Geography: Society, Space and Social Science.* Minneapolis: Minnesota University Press.

Pick, D. (1989). *Faces of Degeneration.* Cambridge, UK: Cambridge University Press.

Pocock, D. C. D. (1981). "Sight and Knowledge." *Transactions of the Institute of British Geographers, NS 6,* 385–393.

Pounds, N. J. G. (1953). *Political Geography.* New York: McGraw-Hill.

Ragin, C. (1987). *The Comparative Method: Moving beyond Qualitative and Quantitative Strategies.* Berkeley and Los Angeles: University of California Press.

Relph, E. (1976). *Place and Placelessness.* London: Pion.

Robbins, P. (2004). *Political Ecology.* Oxford, UK: Blackwell.

Robbins, P., and Fraser, A. (2003). "A Forest of Contradictions: Producing the Landscapes of the Scottish Highlands." *Antipode, 35*(1), 95–118.

Robinson, A. H., Lindberg, J. B., and Brinkman, L. W. (1961). "A Correlation and Regression Analysis Applied to Rural Farm Population in the Great Plains." *Annals of the Association of American Geographers, 51,* 211–221.

Robson, B. T. (1969). *Urban Analysis.* Cambridge, UK: Cambridge University Press.

Rodgers, A. (1952). "Industrial Inertia: A Major Factor in the Location of the Iron and Steel Industry in the United States." *Geographical Review, 42,* 46–56.

Roepke, H. G. (1956). *Movements of the British Iron and Steel Industry–1720 to 1921.* Urbana: University of Illinois Press.

Rose, D. (1987). "Home Ownership, Subsistence and Historical Change: The Mining District of West Cornwall in the Late Nineteenth Century." In N. Thrift and P. Williams (Eds.), *Class and Space.* London: Routledge.

Rostow, W. W. (1960). *The Stages of Economic Growth: A Non-Communist Manifesto.* Cambridge, UK: Cambridge University Press.

Roweis, S. T., and Scott, A. J. (1978). "The Urban Land Question." Chapter 2 in K. R. Cox (Ed.), *Urbanization and Conflict in Market Societies.* Chicago: Maaroufa Press.

Rushton, G. (1969). "Analysis of Spatial Behavior by Revealed Preference." *Annals of the Association of American Geographers, 59,* 391–400.

Russett, B. M. (1964). *World Handbook of Political and Social Indicators.* New Haven, CT: Yale University Press.

Sachs, J. (2003, June). "Institutions Matter, But Not for Everything." *Finance and Development,* 38–41.

Sachs, J., Mellinger, A. D., and Gallup, J. L. (2001). "The Geography of Poverty and Wealth." *Scientific American, 284*(3), 70–75.

Sack, R. D. (1972). "Geography, Geometry and Explanation." *Annals of the Association of American Geographers, 62,* 61–78.

Sassen, S. (2008). *Territory, Authority and Rights: From Medieval to Global Assemblages.* Princeton: Princeton University Press.

Sauer, C. O. (1925). "The Morphology of Landscape." *University of California Publications in Geography, 2,* 19–53.

Sauer, C. O. (1952). *Agricultural Origins and Dispersals*. Washington, DC: American Geographical Society.

Sauer, C. O. (1967). "Agency of Man on Earth." Chapter 1 in F. E. Dohrs and L. M. Summers (Eds.), *Cultural Geography: Selected Readings*. New York: Thomas Crowell. Reprinted from W. L. Thomas (Ed.), *Man's Role in Changing the Face of the Earth*. Chicago: Chicago University Press 1956.

Sayer, A. (1979). "Epistemology and Conceptions of People and Nature in Geography." *Geoforum, 10*(1), 19–43.

Sayer, A. (1984). *Method in Social Science: A Realist Approach*. London: Routledge.

Sayer, A. (1985). "The Difference That Space Makes." Chapter 4 in D. Gregory and J. Urry (Eds.), *Social Relations and Spatial Structures*. London: Macmillan.

Sayer, A. (1989a). "On the Dialogue between Humanism and Historical Materialism in Geography." Chapter 11 in A. Kobayashi and S. Mackenzie (Eds.), *Remaking Human Geography*. London: Unwin Hyman.

Sayer, A. (1989b). "Postfordism in Question." *International Journal of Urban and Regional Research, 13*, 666–696.

Sayer, A. (1992). *Method in Social Science: A Realist View* (2nd ed.). London and New York: Routledge.

Sayer, A. (2004). "Seeking the Geographies of Power." *Economy and Society, 33*(2), 255–270.

Sayer, A., and Walker, R. (1992). *The New Social Economy*. Oxford, UK: Blackwell.

Schaefer, F. K. (1953). "Exceptionalism in Geography: A Methodological Examination." *Annals of the Association of American Geographers, 43*, 226–249.

Schoenberger, E. (2001). "Interdisciplinarity and Social Power." *Progress in Human Geography, 25*(3), 365–382.

Scott, A. J. (1980). *The Urban Land Nexus and the State*. London: Pion.

Scott, A. J. (1982a). "The Meaning and Social Origins of Discourse on the Spatial Foundations of Society." In P. R. Gould and G. Olsson (Eds.), *A Search for Common Ground*. London: Pion.

Scott, A. J. (1982b). "Production System Dynamics and Metropolitan Development." *Annals of the Association of American Geographers, 72*, 185–200.

Scott, A. J. (1985). "Location Processes, Urbanization, and Territorial Development: An Exploratory Essay." *Environment and Planning A, 17*, 479–501.

Scott, A. J. (1988). "Flexible Production Systems and Regional Development: The Rise of New Industrial Spaces in North America and Western Europe." *International Journal of Urban and Regional Research, 12*(2), 171–186.

Scott, A. J. (2000). "Economic Geography: The Great Half-Century." *Cambridge Journal of Economics, 24*, 483–504.

Scott, A. J. (2004). "A Perspective of Economic Geography." *Journal of Economic Geography, 4*, 479–499.

Scott, J. C. (2009). *The Art of Not Being Governed*. New Haven, CT: Yale University Press.

Sealy, K. (1957). *The Geography of Air Transport*. London: Hutchinson.

Seidman, S. (1992). "Postmodern Social Theory as Narrative with a Moral Intent." Chapter 2 in S. Seidman and D. Wagner (Eds.), *Postmodernism and Social Theory*. Oxford, UK: Blackwell.

Semmel, B. (1960). *Imperialism and Social Reform*. Cambridge, MA: Harvard University Press.

Silver, B. (2004). "Labor, War and World Politics: Contemporary Dynamics in World-Historical Perspective." In B. Unfried, M. van der Linden, and C. Schindler (Eds.), *Labor and New Social Movements in a Globalizing World System*. Leipzig, Germany: Akademische Verlagsanstalt.

Silvey, R. (2003). "Spaces of Protest: Gendered Migration, Social Networks, and Labor Activism in West Java, Indonesia." *Political Geography, 22*(2), 129–155.

Smith, D. M. (1977). *Human Geography: A Welfare Approach*. London: Edward Arnold.

Smith, D. M. (1979). *Where the Grass Is Greener: Living in an Unequal World*. London: Penguin.

Smith, D. M. (1988). *Geography, Inequality and Society*. Cambridge, UK: Cambridge University Press.

Smith, J. R. (1913). *Industrial and Commercial Geography*. New York: Henry Holt.

Smith, M. P. (2001). *Transnational Urbanism*. Oxford, UK: Blackwell.

Smith, N. (1984). *Uneven Development*. Oxford, UK: Blackwell.

Smith, N. (1987). "'Academic War over the Field of Geography': The Elimination of Geography at Harvard, 1947–51." *Annals of the Association of American Geographers, 77*(2), 155–172.

Smith, N. (1992). "Contours of a Spatialized Politics: Homeless Vehicles and the Production of Geographical Space." *Social Text, 33*, 54–81.

Smith, N. (2003). *American Empire*. Berkeley and Los Angeles: University of California Press.

Smith, R. J. T., Taaffe, E. J., and King, L. J. (1968). *Readings in Economic Geography*. Chicago: Rand McNally.

Smith, S. J. (1993). "Bounding the Borders: Claiming Space and Making Place in Rural Scotland." *Transactions of the Institute of British Geographers, NS 18*(3), 291–308.

Smith, W. (1949). *An Economic Geography of Great Britain*. New York: Dutton.

Smith, W. (1955). "The Location of Industry." *Transactions of the Institute of British Geographers, 21*, 1–18.

Soja, E. W. (1985). "The Spatiality of Social Life: Towards a Transformative Retheorization." Chapter 6 in D. Gregory and J. Urry (Eds.), *Social Relations and Spatial Structures*. London: Macmillan.

Soja, E. W. (1989). *Postmodern Geographies*. London: Verso.

Sparke, M. (2009). "Triangulating Globalization." *Journal of Historical Geography, 35*, 376–381.

Spencer, J. E., and Horvath, R. J. (1963). "How Does an Agricultural Region Originate?." *Annals of the Association of American Geographers, 53*(1), 74–92.

Springer, S. (2011). "Violence Sits in Places? Cultural Practice, Neoliberal Rationalism, and Virulent Imaginative Geographies." *Political Geography, 30*, 90–98.

Stamp, L. D. (1946). *Britain's Structure and Scenery*. London: Collins.

Stamp, L. D. (1960). *Applied Geography*. Harmondsworth, Middlesex, UK: Penguin.

Stedman Jones, G. (1972). "The History of US Imperialism." Chapter 10 in R. Blackburn (Ed.), *Ideology in Social Science*. London: Fontana.

Stoddart, D. (1989). "A Hundred Years of Geography at Cambridge." *Geographical Journal, 155*(1), 24–32.

Storper, M. (1987). "The Post-Enlightenment Challenge to Marxist Urban Studies." *Environment and Planning D, 5*(4), 418–426.

Storper, M. (1992). "The Limits to Globalization: Technology Districts and International Trade." *Economic Geography, 68*(1), 60–93.

Storper, M. (1997a). *The Regional World.* New York: Guilford Press.

Storper, M. (1997b). "Territories, Flows and Hierarchies in the Global Economy." Chapter 1 in K. R. Cox (Ed.), *Spaces of Globalization.* New York: Guilford Press.

Storper, M., and Walker, R. (1989). *The Capitalist Imperative.* Oxford, UK: Blackwell.

Swyngedouw, E. (1997). "Neither Global nor Local: 'Glocalization' and the Politics of Scale." Chapter 6 in K. R. Cox (Ed.), *Spaces of Globalization.* New York: Guilford Press.

Swyngedouw, E. (2007). "Technonatural Revolutions: The Scalar Politics of Franco's Hydro-Social Dream for Spain, 1939–1975." *Transactions of the Institute of British Geographers, NS 32*(1), 9–28.

Taaffe, E. J. (1974). "The Spatial View in Context." *Annals of the Association of American Geographers, 64,* 1–16.

Taylor, P. J. (1971). "Distances within Shapes: An Introduction to a Family of Finite Frequency Distributions." *Geografiska Annaler, Series B, Human Geography, 53*(1), 40–53.

Taylor, P. J. (1976). "An Interpretation of the Quantification Debate in British Geography." *Transactions of the Institute of British Geographers, NS 1,* 129–142.

Taylor, P. J. (1981). "A Materialist Framework for Political Geography." *Transactions of the Institute of British Geographers, 7,* 15–34.

Taylor, P. J. (1985). "The Value of a Geographical Perspective." Chapter 4 in R. J. Johnston (Ed.), *The Future of Geography.* London: Edward Arnold.

Taylor, P. J. (1989). "The Error of Developmentalism in Human Geography." Chapter 5 in D. Gregory and R. Walford (Eds.), *Horizons in Human Geography.* Totowa, NJ: Barnes & Noble.

Taylor, P. J., and Flint, C. (2007). *Political Geography: World Economy, Nation State and Locality.* Englewood Cliffs, NJ: Prentice Hall.

Taylor, P. J., and Johnston, R. J. (1979). *Geography of Elections.* Harmondsworth, Middlesex, UK: Penguin.

Thomas, E. N. (1960). "Maps of Residuals from Regression: Their Characteristics and Uses in Geographic Research." *University of Iowa, Department of Geography, Report, 2.*

Thompson, E. P. (1967). "Time, Work-Discipline and Industrial Capitalism." *Past and Present, 38*(1), 56–97.

Thompson, E. P. (1971). "Review: Anthropology and the Discipline of Historical Context." *Midland History, 1*(3), 41–55.

Thrift, N. (1983). "On the Determination of Action in Space and Time." *Environment and Planning D, 1*(1), 23–56.

Thrift, N. (1987). "No Perfect Symmetry." *Environment and Planning D, 5*(4), 400–407.

Thrift, N. (1994). "Taking Aim at the Heart of the Region." Chapter 8 in D. Gregory, R. Martin, and G. Smith (Eds.), *Human Geography: Society, Space and Social Science.* Pp. 200–231. Minneapolis: University of Minnesota Press.

Thrift, N. (1999). "Steps to an Ecology of Place." Chapter 15 in D. Massey, J. Allen, and P. Sarre (Eds.), *Human Geography Today*. Pp. 295–322. Cambridge, UK: Polity Press.

Thrift, N., and Williams, P. (Eds.) (1987). *Class and Space: : The Making of Urban Society*. London: Routledge.

Tilly, C. (1984). *Big Structures, Large Processes, Huge Comparisons*. New York: Russell Sage Foundation.

Tobler, W. R. (1965). "Computation of the Correspondence of Geographical Patterns." *Papers and Proceedings of the Regional Science Association, 15*, 131–142.

Tobler, W. R. (1966). *Numerical Map Generalization and Notes on the Analysis of Geographical Distributions* (Discussion Paper No. 8). Ann Arbor: Michigan Inter-University Community of Mathematical Geographers.

Tobler, W. R. (1970). "A Computer Movie Simulating Urban Growth in the Detroit Region." *Economic Geography, 46*, 234–240.

Tobler, W. R., and Wineburg, S. (1971). "A Cappadocian Speculation." *Nature, 231*, 40–41.

Trueman, A. E. (1949). *Geology and Scenery in England and Wales*. Harmondsworth, Middlesex: Penguin.

Tuan, Y.-F. (1971). "Geography, Phenomenology and the Study of Human Nature." *Canadian Geographer, 15*(3), 181–192.

Tuathail, G. (1992). "Political Geographers of the Past: VIII. Putting Mackinder in His Place: Material Transformations and Myth." *Political Geography, 11*(1), 100–118.

Tuathail, G. (1996). *Critical Geopolitics*. Minneapolis: University of Minnesota Press.

Tuathail, G., and Luke, T. W. (1994). "Present at the (Dis)integration: Deterritorialization and Reterritorialization in the New Wor(l)d Order." *Annals of the Association of American Geographers, 84*(3), 381–398.

Tukey, J. W. (1977). *Exploratory Data Analysis*. Pearson.

Ullman, E. L. (1941). "A Theory of Location for Cities." *American Journal of Sociology, 46*, 853–864.

Ullman, E. L. (1956). "The Role of Transportation and the Bases for Interaction." Pp. 862–880 in W. L. Thomas (Ed.), *Man's Role in Changing the Face of the Earth*. Chicago: University of Chicago Press, 1956.

Unwin, T. (1992). *The Place of Geography*. Harlow, Essex, UK: Longman.

Walker, R. A. (1978a). "The Transformation of Urban Structure in the Nineteenth Century and the Beginnings of Suburbanization." Chapter 8 in K. R. Cox (Ed.), *Urbanization and Conflict in Market Societies*. Chicago: Maaroufa.

Walker, R. A. (1978b). "Two Sources of Uneven Development under Advanced Capitalism: Spatial Differentiation and Capital Mobility." *Review of Radical Political Economy, 10*(3), 28–39.

Walker, R. A. (1981). "A Theory of Suburbanization: Capitalism and the Construction of Urban Space in the United States." Pp. 383–429 in M. J. Dear and A. J. Scott (Eds.), *Urbanization and Urban Planning in Capitalist Society*. London: Methuen.

Walker, R. A. (1985). "Class, Division of Labor and Employment in Space." Chapter 8 in D. Gregory and J. Urry (Eds.), *Social Relations and Spatial Structures*. Pp. 164–189. Basingstoke: Macmillan.

Warde, A. (1988). "Industrial Restructuring, Local Politics and the Reproduction of Labor Power: Some Theoretical Considerations." *Environment and Planning D, 6*(1), 75–95.

Warren, C. R. (2007). "Perspectives on the 'Alien' versus 'Native' Species Debate: A Critique of Concepts, Language and Practice." *Progress in Human Geography, 31*(4), 427–446.

Watson, J. W. (1955). "Geography: A Discipline in Distance." *Scottish Geographical Magazine, 71*, 1–13.

Watts, M. J. (1983a). "Hazards and Crises: A Political Economy of Drought and Famine in Northern Nigeria." *Antipode, 15*(1), 24–34.

Watts, M. J. (1983b). "On the Poverty of Theory: Natural Hazards Research in Context." Pp. 231–262 in K. Hewitt (Ed.), *Interpretations of Calamity*. Boston: Allen and Unwin.

Watts, M. J. (1983c). *Silent Violence*. Berkeley and Los Angeles: University of California Press.

Werner, C. (1968). "The Role of Topology and Geometry in Optimal Network Design." *Papers and Proceedings of the Regional Science Association, 21*(1), 173–189.

Western, J. S. (1978). "Knowing One's Place: 'The Colored People' and the Group Areas Act in Cape Town." Pp. 297–318 in D. Ley and M. Samuels (Eds.), *Humanistic Geography*. Chicago: Maaroufa.

Western, J. (1981). *Outcast Cape Town*. Berkeley and Los Angeles: University of California Press.

Wise, M. (1949). "On the Evolution of the Jewellery and Gun Quarters in Birmingham." *Transactions of the Institute of British Geographers, 15*, 57–72.

Wolfe, A. (1981). *America's Impasse: The Rise and Fall of the Politics of Growth*. New York: Pantheon Books.

Wolpert, J. (1964). "The Decision Process in Spatial Context." *Annals of the Association of American Geographers, 54*, 537–558.

Wolpert, J. (1965). "Spatial Aspects of the Decision to Migrate." *Papers and Proceedings of the Regional Science Association, 15*, 159–172.

Wolpert, J. (1970). "Departures from the Usual Environment in Locational Analysis." *Annals of the Association of American Geographers, 50*(2), 220–229.

Wright, E. O. (1983). "Giddens' Critique of Marxism." *New Left Review*, No. 138, 11–35.

Wrigley, E. A. (1965). "Changes in the Philosophy of Geography." Chapter 1 in R. J. Chorley and P. Haggett (Eds.), *Frontiers in Geographical Teaching*. London: Methuen.

Wylie, J. (2006). "Poststructuralist Theories, Critical Methods and Experimentation." Chapter 27 in S. Aitken and G. Valentine (Eds.), *Approaches to Human Geography*. London: Sage.

Wylie, J. (2007). *Landscape*. London: Routledge.

Yeates, M. (1963). "Hinterland Delimitation: A Distance-Minimizing Approach." *Professional Geographer, 15*(6), 7–10.

Yeates, M. (1974). *An Introduction to Quantitative Analysis in Human Geography*. New York: McGraw-Hill.

Index

About the Author

Kevin R. Cox, PhD, is Distinguished University Professor of Geography at The Ohio State University and has been a Guggenheim Fellow. He has written a number of books, including *Political Geography: Territory, State and Society* and *Spaces of Globalization: Reasserting the Power of the Local.*